~LIVING~ BY WATER

Essays on Life, Land & Spirit

~LIVING~
BY WATER

Essays on Life, Land & Spirit

BRENDA PETERSON

Alaska Northwest Books™
Anchorage • Seattle

Acknowledgments

I would like to gratefully acknowledge the help and encouragement of
Susan Biskeborn, J. Kingston Pierce, and my editor, Marlene Blessing.

The essays in this book first appeared in the following publications,
in slightly different form:
"Watching for Whales in Winter," "Shadow People," and "Wild, for
All the World to See" in *Washington* magazine; "Growing Up Game"
in the *Seattle Weekly* and *Graywolf Annual Three* (1986); "Oil Spill
Eulogy" in the *Seattle Weekly*; "Believing the Bond" (under the title
"The Family Tree") and "Where the Green River Meets the Amazon"
in *Pacific Northwest* magazine; "Living by Water" in *Seattle Home and
Garden*.

Library of Congress Cataloging-in-Publication Data
Peterson, Brenda.
    Living by water : essays on life, land, & spirit. / by Brenda Peterson.
      p.   cm.
  ISBN 0-88240-358-3 — ISBN 0-88240-400-8 (pbk.)
  1. Aquatic biology. 2. Natural history. 3. Ecology. I. Title.
  QH90.1.P47 1990
  508 — dc20   90-36821
                                      CIP

Designed by Alice Merrill Brown
Edited by Marlene Blessing
Cover art is a detail from a painting by Cecilio Sanchez,
    entitled *Fundo Marino*.

Alaska Northwest Books™
A division of GTE Discovery Publications, Inc.
22026 20th Avenue S.E.
Bothell, Washington 98021

Printed in U.S.A.

To Puget Sound, who mothers me

Contents

Shadow People

I moved here from many states. First, from the desert mesas of Arizona, where the Hopi Indians believe they keep the whole world spinning by their prayers and rituals. I also journeyed here from New York City, where what keeps the world spinning seems man-made. (Once during the great power failure of the late seventies, a co-worker woke up, shocked to find the sun still rising.) I also moved to this Northwest water, fog, stillness, and rain from the equally mist-shrouded South, where fundamentalists as fervent and ever-present as kudzu eagerly await the Second Coming — that stopping of the world's spin — with rapture. Nine years I have stayed here in Seattle, the longest I've ever lived anywhere, because I've come to suspect that this is a shamanistic state.

The word *shaman* has its origin in the Turko-Tungusian Siberian language and describes those healers, often wounded themselves, who undertake journeys to enter the spirit world, there to witness, experience, and consciously bring back a vision for the tribe or what Indians call simply "the People." The People is also what Northwest Indians called animals, who existed long before humans and who were mythic giants, gods, and goddesses. Everyone knows that Coyote, or Changer, that wily and benevolent trickster, helped create the human tribes of the Northwest; but he had guidance from all-seeing Raven, that high-flying Eagle spirit, from the not-to-be-outdone Fox, and from clever Blue Jay.

Whenever we forget these first stories, and so fall sick, out of sync, the shaman in each of us can remember and perhaps heal ourselves here in our homeland. This remembering of our proper place among the animals and the land, this inward-turning journey to connect the visible world with the invisible one that parallels our daily lives, is very much a part of our Northwest heritage and future. It also may well be how we gauge our success and the success of others.

Success in, say, Los Angeles, where the sun always shines, or New York, where city dwellers equate the sun with Con Edison, is a much more external affair. Perhaps if there is always illumination from lights or the sun, one must learn to live in the spotlight, seeking the Right Act, the Right Clothes, the Right Car, the Right Relationships. But in the misty San Juan Islands, in the gray, rain-swept cityscapes of Seattle, in the high, snowed-in hollows of the

Olympics, and the cloud-shadowed deserts of Eastern Washington there is so much hidden about the land, the lives. Here, what you see is not what you get. Here the Biblical scripture holds more true, for everywhere is "the evidence of things not seen."

It is this mysterious evidence that Northwesterners have learned to value. It is also what lends us a mystical sense, both of our inner and outer landscapes. Look at Mount Rainier. She is seen and she is not seen, but that most mesmerizing of mountains is always present and felt. I've often told the story to my East Coast friends of my first year in Seattle. From January through March I'd not once seen "the Mountain." I'd heard Rainier called this with the reverence usually reserved for great spiritual teachers, visionary leaders, artists.

Privately I scoffed at such deference to only one mountain — I, born in the High Sierra among a whole tribe of massive snow-shrouded mountains.

But then one day I was driving across the Mercer Island bridge and casually looked southward. What rose up over the water was nothing less than a blazing goddess. I came close to dying at that moment, not from awe, but because I literally lost control of the steering wheel. A friend righted it for me.

I still count that moment — meeting what I can only call a divine mother of a mountain — as a fixed point by which I navigate my life. For just as Mount Rainier broke through her months of mist, so it seemed to me that another world shone there on the horizon — the healing mystical world stood revealed beside my daily life. It reminded me of what Black Elk said, that "the central mountain of the world is where you are."

Many Indian stories begin, "Long ago when mountains were people . . ." The Salish tale of Mount Rainier (also known as Tahoma, or "the great mountain, which gives thunder and lightning, having great unseen powers") tells of how this earth goddess Tahoma swallowed up land, tribes, animals and, finally, Changer, who was disguised as Fox. Tahoma gobbled up so much rock and water that she at last burst open, lavalike blood flowing down her sides. Changer made her into the translucent volcanic presence that Washingtonians today revere.

What we worship about where we live is what we decide is successful. Shamanism is not really a form of worship, it is an attitude toward life and all that is living. A mystic's path is to hold dear shamanism's nondogmatic, direct link with the spirit world as seen in the deities of land, water, and animals. This is quite different from the hierarchical class system of the East Coast, where what is holy is most often man-made. What we in the West strive to balance is the beautiful in both nature and art. We've had the advantage of learning from the environmental ravages of the East; we also live in the shadow of our own Western ancestors who all but erased an entire Indian culture, as well as their sacred buffalo, in the span of a single decade.

This heritage lets us come full circle in our mythology. We've moved from the what-you-see-is-what-you-get to what-you-can't-see-is-what-you-must-get inner journey of the shaman.

If on the East Coast the journey was west across a continent to claim land, on the West Coast we have nowhere to go but across water. As everyone knows (and Eagle will teach you), humans can't claim the sea. She comes and goes like Mount Rainier; she is too deep to till or plunder; she eludes our dominion just as she does our myth of ownership.

Not forgetting our balance and connection with the land and water, not forsaking our inward journey to remember a time when we were ourselves animals, mountains, and even the fertile salt water that first spawned us, is what the Northwest mysticism is all about. The shaman heals him- or herself, not by an external, established authority, not in battle to conquer nature, but by changing on the inside, telling the story and so, perhaps, helping to heal the world.

There is an everydayness about this, a grounded and conscious direction, a sense of humor and self-mastery that seems Oriental. In the Northwest, where there is a mingling of Asians, American Indians, and white settlers who live by water, this mystical tradition looks to have found a spiritual territory.

Tess Gallagher, the native Port Angeles poet, said in a recent interview,

> In the East, you take your measure more often from the community of man, but here you're constantly aware of a very different kind of space and time. . . . I watch the tides come in and go out. I know there's a force at work that's very mysterious and which is doing its thing, all the time I think I'm doing my very important thing. . . . Part of what you're doing as a writer is to make that silent language of mountains and trees and water part of your language. It's speaking all the time and I hear it speaking.

This aliveness of all things is the most sacred element of a shamanistic state. If we know everything is living as we are, then we can hear the mischievous voice of Coyote cajoling us, "We animals were first and we created your tribes — listen to your elders!" We might also truly listen to the music Puget Sound makes, teaching us to be still and hear the deep within. We might see how slightly we know ourselves if we don't also learn to love the shadows inside the gray and healing mists of our Northwest skies.

Growing Up Game

*W*hen I went off to college, my father gave
me, as part of my tuition, fifty pounds of moose meat. In
1969, eating moose meat at the University of California was
a contradiction in terms. Hippies didn't hunt. I lived in a
rambling Victorian house that boasted sweeping circular
staircases, built-in lofts, and a landlady who dreamed of
opening her own health food restaurant. I told my housemates that my moose meat in its nondescript white butcher
paper was from a side of beef my father had bought. The

carnivores in the house helped me finish off such suppers as sweet-and-sour moose meatballs, mooseburgers (garnished with the obligatory avocado and sprouts), and mooseghetti. The same dinner guests who remarked upon the lean sweetness of the meat would have recoiled if I'd told them the not-so-simple truth: that I grew up on game, and the moose they were eating had been brought down, with one shot through his magnificent heart, by my father — a man who had hunted all his life and all of mine.

One of my earliest memories is of crawling across the vast continent of crinkled linoleum in our Forest Service cabin kitchen down splintered back steps, through wild-flowers growing wheat-high. I was eye-level with grasshoppers who scolded me on my first solo trip outside. I made it to the shed, a cool and comfortingly square shelter that held phantasmagoric metal parts; they smelled good, like dirt and grease. I had played a long time in this shed before some maternal shriek made me lift up on my haunches to listen to those urgent, possessive sounds that were my name. Rearing up, my head bumped into something hanging in the dark; gleaming white, it felt sleek and cold against my cheek. Its smell was dense and musty and not unlike the slabs of my grandmother's great arms after her cool, evening sponge baths. In that shed I looked up and saw the flensed body of a doe; it swung gently, slapping my face. I felt then as I do even now when eating game: horror and awe and kinship.

Growing up those first years on a Plumas National Forest station high in the Sierra Nevada near Oregon was somewhat like belonging to a white tribe. The men hiked off every day into their forest and the women stayed behind in the circle of official cabins, breeding. So far away from a store, we ate venison and squirrel, rattlesnake and duck. My first rattle, in fact, was from a diamondback rattler my father killed as we watched, by snatching it up with a stick and winding it, whiplike, around a redwood sapling. Rattlesnake tastes just like chicken, but has many fragile bones to slither one's way through. We also ate rainbow trout, rabbit, and geese galore. The game was accompanied by such daily garden dainties as fried okra, mustard greens, corn fritters, wilted lettuce (our favorite because of that rare, blackened bacon), new potatoes and peas, stewed tomatoes, barbecued butter beans.

I was four before I ever had a beef hamburger, and I remember being disappointed by its fatty, nothing taste and the way it fell apart at the seams whenever my teeth sank into it. Smoked pork shoulder came much later, in the South; and I was twenty-one, living in New York City, before I ever tasted leg of lamb. I approached that glazed rack of meat with a certain guilty self-consciousness, as if I unfairly stalked those sweet-tempered white creatures myself. But how would I explain my squeamishness to those urban sophisticates? How explain that I was shy with mutton when I had been bred on wild things?

Part of it, I suspect, had to do with the belief I'd also been bred on: we become the spirit and body of animals we eat. As a child eating venison, I liked to think of myself as lean and lovely just like the deer. I would never be caught dead just grazing while some man who wasn't even a skillful hunter crept up and conked me over the head. If someone wanted to hunt me, he must be wily and outwitting. He must earn me.

My father had also taught us as children that animals were our brothers and sisters under their skin. They died so that we might live. And of this sacrifice we must be mindful. "God make us grateful for what we are about to receive," took on new meaning when we imagined the animal's surrender to our own appetites. We also used *all* the animal, so that an elk became elk steaks, stew, salami, and sausage. His head and horns went on the wall to watch us more earnestly than any baby-sitter, and every Christmas Eve we had a ceremony of making our own moccasins for the new year out of whatever Father had tanned. "Nothing wasted," my father would always say, or, as we munched on sausage cookies made from moosemeat or venison, "Think about who you're eating." We thought of ourselves as intricately linked to the food chain. We knew, for example, that a forest fire meant, at the end of the line, we'd suffer too. We'd have buck stew instead of venison steak, and the meat would be stringy, withered-tasting, because in the animal kingdom, as it seemed with humans, only the meanest and leanest and orneriest survived losing their forests.

Once when I was in my early teens, I went along on a hunting trip as the "main cook and bottle-washer," though I don't remember any bottles; none of these hunters drank alcohol. There was something else coursing through their veins as they rose long before dawn and disappeared, returning to my little camp most often dragging a doe or pheasant or rabbit. We ate innumerable cornmeal-fried fish, had rabbit stew seasoned only with blood and black pepper.

This hunting trip was the first time I remember eating game as a conscious act. My father and Buddy Earl shot a big doe and she lay with me in the back of the tarp-draped station wagon all the way home. It was not the smell I minded, it was the glazed great, dark eyes and the way that head flopped around crazily on what I knew was once a graceful neck. I found myself petting this doe, murmuring all those graces we'd been taught long ago as children. *Thank you for the sacrifice, thank you for letting us be like you so that we can grow up strong as game.* But there was an uneasiness in me that night as I bounced along in the back of the car with the deer.

What was uneasy is still uneasy — perhaps it always will be. It's not easy when one really starts thinking about all this: the eating game, the food chain, the sacrifice of one for the other. It's never easy when one begins to think about one's most basic actions, like eating. Like becoming what one eats: lean and lovely and mortal.

Why should it be that the purchase of meat at a butcher shop is somehow more righteous than eating something wild? Perhaps it has to do with our collective unconscious that sees the animal bred for slaughter as doomed. But that wild doe or moose might make it without the hunter. Perhaps on this primitive level of archetype and unconscious knowing we even believe that what's wild lives forever.

My father once told this story around a hunting campfire. His own father, who raised cattle during the Great Depression on a dirt farm in the Ozarks, once fell on such hard times that he had to butcher the pet lamb for supper. My father, bred on game or their own hogs all his life, took one look at the family pet on that meat platter and pushed his plate away. His siblings followed suit. To hear my grandfather tell it, it was the funniest thing he'd ever seen. "They just couldn't eat Bo-Peep," Grandfather said. And to hear my father tell it years later around that campfire, it was funny, but I saw for the first time his sadness. And I realized that eating had become a conscious act for him that day at the dinner table when Bo-Peep offered herself up.

Now when someone offers me game, I will eat it with all the qualms and memories and reverence with which I grew up eating it. And I think I will always have this feeling of horror and awe and kinship. And something else — full knowledge of what I do, what I become.

Animals as Brothers and Sisters

*A*s a child I played a game with my siblings: *What country are you? What body of water? What war? What animal?* My sister was Ireland, the South Seas, the War of Independence, and a white stallion. My brother was Timbuktu, the Amazon River, the One Hundred Years War, and a cobra. I was South America, the Gulf of Mexico, the Civil War, and a dolphin. Sometimes we called upon our animals — my sister galloping away from grown-ups with a powerful snort and a flick of her fine, silver mane; my brother summoning the fierce serpent hiss to ward off his older sisters; and I, soaring through sea and air with my tribe of dolphins.

Our parents didn't think it odd then that their children metamorphosed into animals, oceans, or wars right there in the middle of the living room or backyard. My father always planted his family next to a forest, a river, or an ocean — all of which were expansive and natural enough to absorb our wildest play. One of the few times our transformation was curbed was at the dinner table — if, say, my brother as cobra poised above my hand as I cut the cake in exact equal pieces or if my sister was pawing the tablecloth with her pale equine impatience. Then my father, whose own play was raising horses and hunting, might threaten my sister with a tight bit or suggest my brother uncoil himself and cool down until his blood was really reptilian, slow and grounded.

"The cobra can't uncoil until he strikes and eats," my brother would mutter as he sighed and right before us changed back into the youngest child. But his eyes remained hooded.

"The white stallion is never broken," my sister would warn my father, who did raise her with a freer hand as if she were one of his fine, high-strung thoroughbreds.

I was always under water during these discussions, on the green, shady depths of my warm gulf, listening more intently to a language that creaked and chattered like high-speed hinges — dolphin gossip. Or sometimes I just tuned in to their other language: the pictures dolphins send one another in their minds. Because I had to come up for air, and my eyes were as good above water as below, I did keep a lookout on my family's dinner dramas. But if my mother was having one of her bad moods or my father was giving his lectures, back down I'd go to my other family, who welcomed me with wide-open fins. Even without hands, the dolphins embraced me more than most people did. It was body-to-body, full embrace, our eyes unblinking, utterly open as we swam, belly-to-belly, our skin twenty times more sensitive than that of humans.

The play my siblings and I chose as children is mirrored in the way we live as grown-ups. And I suspect it has much to do with our career choices, our relationships, even where we choose to live. My sister finds her South Seas body of water (and reunites with our family's Seminole blood) by living in Florida and marrying into an old Key West family. She is still fighting her War of Independence, a ripsnorting battle, which involves her husband and three daughters as high-spirited playmates. Every so often I see her snort and toss her full mane of hair; and when she really means business, she paws the ground with her delicate, high-heeled hooves. My brother, as a Navy jet navigator, has traveled the world, is caught up in all sorts of military intrigue in far places — enough to last one hundred years easy. His serpentine ways have surrendered more to the feminine aspects of the snake, for at the births of his two daughters, my brother shed his toughened military skin and was reborn. And me, well I now live in a whole city under water: Seattle. And I'm still swimming with dolphins.

This is difficult to do in Puget Sound with its year-round temperature variation from forty-six to forty-eight degrees Fahrenheit. So aside from sighting dolphin schools from shore or ferry, I've had to go to warmer waters to make my psychic life match up with my actual life. How convenient then that my sister's Conch Republic connections carried me to the Florida Keys to find my animal allies.

Actually, it was a kind of coincidence. Four years ago I was sitting in my Seattle study listening to the splatter of rain on my roof, reading a *New York Times* article about a Florida Keys research program that reversed our society's usual prejudice against animals: the dolphins were not there for human amusement. Instead, we were their "toys," and the researchers studied the interaction between humans and dolphins while in the cetacean's own environment. Everything was geared toward what fascinated the dolphins, what made them choose a particular person as a swimming partner over another. Researchers don't know why, but dolphins prefer children first, then women, and then men. But why do they ignore some people completely and gather around others with absolute attention?

As I was reading this article, my sister Paula called. Seems she was stranded in a motel along the string of coral keys, en route to Key West. I could just hear her champing at the bit. "We're stuck here overnight. The girls are bored silly," she said. They weren't the only ones, I suggested, then told her about the dolphin research that just happened to be only miles from their motel.

"All right, all right, we'll go swim with your dolphins," Paula said.

My sister was eight months pregnant with her third daughter, and none of her deliveries had been simple. That's why my sister, lowering herself and her swollen belly into the warm tropical water, showed her true mettle — she was, as my father always recognized, fine, great-hearted horseflesh. My nieces needed no courage to dive into the lagoons where dolphins chattered about them. It was delight at first sight.

These dolphins are in "elective captivity," which means the underwater fences that separate them from the saltwater canal leading to the ocean are opened twice a day to let the dolphins return to their home territory. They return to the research center of their own free will. No dolphin has ever chosen to escape; they seem as fascinated with humans as we are with them, though we've given them much reason to keep their distance. There is no record, since antiquity, of a dolphin harming humans; yet we routinely kill thousands of dolphins every year with our tuna industry's tactics of drift-net fishing. In our search for tuna, our nets trap hundreds of dolphins a day. They die dreadfully; they drown.

Because dolphins breathe as we do, nurse their young, and are warm-blooded, there is a mammalian bond, which perhaps explains why dolphins have anything whatsoever to do with humans. The bond was evident as Paula lost her pregnant waddle and floated weightless, waiting for the dolphins. But first they played with her daughters. My four-year-old niece, Lauren, with the fierce grip of all newborns and single-minded children, grabbed hold of a dorsal fin and held on as she was sped around the lagoon at what seemed like the speed of light. She doesn't remember seeing anything but bright bubbles. Careful to keep Lauren's small head above water, her dolphin, who weighed about three hundred pounds and was itself a relative child (only six years old in a lifespan of approximately forty years), carried Lauren as it would a precious baby doll.

Another dolphin swam sister Lindsay, two and a half, round and round until she was dizzy. Then they let her bob about in her life jacket, singing at the top of her lungs. The dolphins showed their approval with some tail slaps, spins, and leaps, always careful about their motors, those great tails. With their phenomenal 360-degree overlapping vision, the dolphins always know exactly where you are. After playing with the children, the dolphins circled my sister, and when their echolocation heard the fetal heartbeat, they got very excited. The high-frequency whines and creaks increased as their sonar sounded my sister's belly, read the fetal blood pressure, and scanned the infant's stomach gases for signs of stress.

"What are they doing?" Paula asked. Her whole body was buzzing.

"Offering to midwife you," the researcher replied. "They seem concerned about the baby. Is . . . is there anything wrong?"

"I don't think so," Paula answered, and for the first time in that lagoon, she felt fear. My sister is a nurse and knows all about ultrasound. But perhaps there was something the dolphins deciphered that our technology didn't.

Then the researcher told my nieces how the dolphins midwife one another, assisting the mother as she swirls and spins in labor by stroking her flanks and at the moment of birth, when the newborn dolphin eases out of one watery womb into another, the midwives lift the calf with their long, sensitive beaks up to the surface. There the newborn dolphin takes its first breath. Every breath thereafter for the rest of the dolphin's life will be taken consciously. A cetacean's brain, somewhat larger than that of a human's, has had thirty million more years of evolutionary development than our species. Some scientists theorize the dolphins exist in an alpha state — what we experience as meditation — and since they never really sleep, just switch sides of the brain being used, researchers wonder what kind of intelligence is here.

My sister certainly wondered when she gave birth three weeks later to a daughter with a rare blood disease. Had the dolphins diagnosed it? After much trauma and weeks of watching her newborn double as a tiny human pincushion, Paula brought her daughter, Lissie, home from the hospital.

On Lissie's second birthday, in gratitude and out of curiosity, we took her back to swim with dolphins. But the rules had changed: no pregnant women and only children who are excellent swimmers. So little Lissie jumped and leaped on the side of the lagoon, shouting, "I am a dolphin! I am a dolphin!" as her sisters and Paula and I all slipped back into the warm salt water.

This was my second swim with dolphins, and my first time at the Dolphins Plus Marine Mammal Research & Education Center in Key Largo. My dolphin companions,

Niki, Dreamer, and Sara, were six-year-old females in elective captivity only two years. Exuberant, still quite wild, they are children themselves.

"Remember," our researcher reminded us as we eased into the water, "you'll have to be creative if you want them to play with you — don't just bob about gawking. They've already got enough float toys."

The dolphins here are not rewarded with food for interaction with humans; that is the old model of performance. Food is given the dolphins at another time, separate from human-cetacean interaction. The real reward for all of us is the play itself.

As I swam, snorkel mask down, arms at my side to signal that I would wait for them to choose to play with me, I heard far below the familiar high-frequency dialogue. It sounded like the high-pitched whine of a jet engine right before takeoff, combined with rapid creaks and bleeps. The sounds encircled my body and then, as the dolphins came closer, there was that astonishing physical sensation of being probed by their sonar. It's as subtle as an X ray, but exhilarating. My whole body tingled, stomach gurgled, head felt pleasurable pricking as if a high-speed ping-pong game played with light was bouncing around my brain.

I am reminded of my friend John Carlyle, who has researched, trained, and played with dolphins for twenty years, telling me of one of his experiments with dolphin echolocation. Trying to discern the limits of dolphin sonar, he placed eyecups on his dolphins and then asked them to recognize certain symbols by echolocation. In an experiment that had taken him months to design, the dolphins learned the symbols in five minutes. So John had to come up with more difficult ways of testing the depth of their sonar. After much research, his final experiment, which was the limit of human technology at the time, discovered dolphins could discern a symbol one one-thousandth of an inch square; they could also differentiate varying carbon densities in metal rods, and distinguish colors — all by echolocation. Knowing of their precision made the experience of having my body echo-scrutinized more than simply a physical sensation. I was scanned more profoundly than by anything our medical science has yet invented. But there is another element here, not at all scientific. It is what happens to my heart, not physiologically, but emotionally.

Every time I'm sounded by a cetacean, I feel as if my cells are penetrated, seen, and — what is most remarkable — *accepted*. I've never felt judgment, even if the dolphin chose to bypass me for another playmate. The Dolphins Plus researchers report that often, whether a dolphin spends five or forty-five minutes with a swimmer, that person will say it was enough, all they needed, as much as they could receive. In fact, every time I've swum with dolphins, my human companions have admitted afterward that we each felt like the favorite. Could it be we have something to learn about parenting from dolphins?

As I swam on the surface, peering through my mask into the dense green depth, I wondered what I must look like to a dolphin. Humans are the most ungainly mammals dolphins see in the ocean. We are the only creatures in the sea who splash at both ends of our bodies. Our appendages don't move in sync with the sea as do the long arms of anemones. There is only one dance in the sea, one pulsing movement of all that lives, one law. Even if one of our bombs exploded here, its harm would be muted. And after it exploded, its metal innards would settle to the sea bottom as no more than an artificial reef adorned by pink brain coral, starfish, and barnacles. Swimming and hoping a dolphin might play with me, I wondered about those ancient dolphin ancestors who decided millions of years ago — while humans were still hanging from trees — to go back to the sea. Did those early cetaceans foresee the fate of our species' self-destruction? Is that why they left us to our weapon-and-tool-making hands (the use of which takes up so much of our brain's functions), while their skeletal hands slowly evolved into flippers to flow *with* rather then change their environment?

The dolphins always come when I'm most distracted, when my mind is not on them at all, but drifting, perhaps dreaming. In my floating reverie, I was startled by the sensuous skin stroking my legs. I happily recognized the silken, clean, elegant feel of dolphin belly as Dreamer ran her whole body across my back like a bow glides across violin strings. And then she was gone. There were only the sounds fading, then coming closer as suddenly all three young dolphins swam toward me. I still see in my dreams those gray globelike domes with brown, unblinking eyes meeting mine as the dolphins greeted me under water. "Intimate" is the only way I can describe their eye contact. Benevolent, familiar, and again that acceptance. Any fear one feels vanishes once those eyes hold yours.

"Choose one!" the researcher shouted above me. Having been under water so long, I could barely hear his voice. But I remembered his instructions; the dolphins are possessive of their toys and I must bond with only one or else they'd squabble among themselves. So I chose Niki, though Dreamer was my favorite, because if truth be told, Niki made the choice. She slipped her dorsal fin under my arm and raced off with me at such a speed I saw only bubbles and sky. Then she dove with me and we both held our breaths. As we surfaced, I saw in the opposite lagoon two dolphins leaping with my nieces like calves in tow. No time to see anything else — I inhaled and dove down again.

Thrilling, this underwater ballet, as I twirled with my dolphin, my hands along her flanks. Fluid, this liquid life below where all is weightless and waves of warmth enfolded my body as I breathed air in this watery element. And I was not alone. Everywhere was sound — my nieces singing, and the dolphins' dialogue. My mind suddenly filled with pictures. Then I realized that every time I imagined my dolphin doing something, a split second later she did it. It was not a performance at my request; it was an answer to my wondering. Call and response. It was also an anticipation of my delight, a *willingness* that is the purest form of play.

I pictured myself spinning round, one hand on Niki's heart — it happened. I saw both my arms outstretched, a dolphin's dorsal offered each hand — and suddenly I was flying between Niki and Dreamer. It was impossible to tell who was sending whom these pictures. But they all happened. It was like instant replay of everything imagined. And now I understood why the child in me chose dolphins. What more perfect playmates?

Ahead in the water swam my sister. Paula was galloping with her dolphin; and my niece Lauren had a dolphin gently resting its long beak on her legs like a paddle to push her through the water. Distracted, I broke one of the basic rules: I got too close to a dolphin and her favorite toy (Paula). Suddenly a wallop to my shoulder. My world turned upside down and though I was face-up in the air, I breathed water. Sputtering, I broke another rule: my body tilted vertically, a sign to the dolphins of distress. Another whack of a pectoral in the lower back, then a beak thrust under my bottom to raise me above the water.

"Horizontal!" the researchers yelled. "They think you're drowning."

I would rather play with a dolphin than be rescued by one. Those whacks are painful reprimands, a lesson in life and death to a wayward human. Blowholes fiercely expelled their air everywhere around me. Surrounded by all three dolphins, I started to cry. I failed, I felt. I was a fool. And for the first time ever, I was afraid of them.

It was hard not to cower there in the water with them.

All the pictures flooding my mind overwhelmed me and I couldn't figure anything out. Except I remembered to float, though my body was rigid and what I would most have liked to do was curl up into a fetal ball and be safe on shore the way long ago I'd surface from my own darker daydreams to find myself at the comfortingly ordinary dinner table I first sought to escape. But this was real; I couldn't imagine my way out of it. Or could I?

Again and again one picture in my head. It was I, still shaken, but surrendering to all three dolphins at once. I breathed raggedly, the snorkel like an intruding fist into my mouth. But after closing my eyes, I allowed it. Yes, they could come back and find me again where I floated in fear. At first Niki and Sara were tentative, their beaks very gently stroking my legs. Now that I wasn't going to drown, would I play with them again?

I am small, I thought, and hoped they could hear. *I am just a human being — afraid and fragile in your element. Be careful with me?*

And they were. Together the three of them floated me so slowly my body barely rippled water. Then began the deepest healing. Dreamer gently eased me away from the others with a nudge of her dorsal fin. Her eyes steadily held mine as she swam gracefully in wide arcs of figure eights around the lagoon. In and out through warm water. My body surrendered to the massage, not of hands, but of water and sound. I thought of the others who come here who are not as healthy as I — the autistic and Down's syndrome children, the handicapped, the terminally ill, all of whom are nursed by the dolphins who embraced me. Deeper than the play, more moving than the sense of another mind in these waters, was the simple kindness of the creatures. I did not understand it. I wanted to.

When I closed my eyes, the pictures grew stronger, as did my senses. My hands slid down Dreamer's silken body, memorizing notches and scars as a blind woman does her loved one. I remembered that in China those born blind were believed to be the most gifted masseurs — because hands are another way of seeing. My hands still hold the exact feel of dolphin skin. Even now, across time and continent, my hands can still grow warmer, tingle with the memory of that cool, sleek skin that trembles when touched.

Dreamer's name comes from her eyes. Half-lidded, there is in her mild, dark eyes a different light. Sloe-eyed, they call it down South — and the sweet, fizzy drink made from those black sloe berries is a euphoric mix reminiscent of humid, fragrant Southern nights. *Down home*, I thought, as we glided through dark green depths. I closed my eyes and felt that this underwater world, too, was down home.

As Dreamer circled with me, I was so relaxed I barely recognized the voices far above in the high, harsher air. "Come back," the researcher called. "It's time . . . but the dolphins are having so much fun with you, they're not going to let you get out easily."

There was a firmness to our researcher's voice, like a parent calling children in from play. We'd been swimming with the dolphins three times longer than the allotted half hour and I suddenly realized I was utterly exhausted. I felt like I'd been moving heavy furniture for days. My snorkel mask fogged and the balmy wind felt abrupt. I remembered gravity and how it works against me.

"You'll all have to link arms to signal the dolphins that you're serious about getting out. They'll respect your tribe. But they'll protest!"

As my sister and my nieces and I moved toward the dock, the dolphins leapt over our heads, chided us for spoiling their sport, and swam figure eights between our legs. Even as we hoisted ourselves up onto the platform, Niki cajoled my niece by opening her long beak and running it up and down Lauren's small leg.

"She's testing you," the researcher told Lauren and laughed. "They only do that when they *really* like you."

"It's a compliment," Lauren confirmed in her most matter-of-fact voice. "She likes me best."

Of course we all secretly felt that way. I still feel that way, these many months later as I sit surrounded by my photos — Dreamer's eyes still hold mine as she glides by in a shining green background; Niki exuberantly leaps above the surface; Sara offers her abiding companionship. My niece Lauren sends me drawings of dolphins and whales; her sister Lindsay has decided to speak nothing but dolphin dialect when I call long-distance. We cluck and click and make sounds deep in our throats like a Geiger counter. Anyone crossing wires on our conversations might think there was electronic equipment trouble on the line.

So the second generation of our family carries on the tradition of claiming our animals within. I often tell my nieces a story about the Northwest Coast Indian belief that before humans came on the scene, the world was made up of beings whose spirits could change into anything from salmon to rock, from raven to waterfall. Every form had its lessons. And the human form, being the most recent, was considered to have the most lessons. Long ago, when my own father taught me that animals were part of our larger family, he echoed this common wisdom. It is wisdom that still serves me well.

For example, I'll often find myself in some situation thinking: "What would a dolphin do?" Animals do not change the world; they adapt. In my own life, the flexibility and adaptability of a dolphin mind, their sense of tribe and play, guides me. I can call upon the dolphin inside me for counsel as well as companionship. And the irony is that apprenticing myself to an animal like a cetacean somehow teaches me more about becoming a human being. As John Lilly, the respected dolphin researcher, wrote, "You see, what I found after twelve years of work with dolphins, is that the limits are not in them, the limits are in us. So I had to go away and find out, who am I? What's this all about?"

These days here in the Northwest, where the Puget Sound offers some of the world's most fertile habitat for marine animals, I realize that becoming a dolphin, like becoming a human, is a state of mind. Here, where the Native Americans remind us that everything shimmers with its own inner life, I walk by the water and send greetings in the form of mental pictures to all the mammals that swim in the Sound. Often I visit the belugas at Point Defiance Zoo, especially when I am sad or in need of inspiration. Those white, generous creatures gaze at me through glass and I put my hands up, hoping they'll feel my own little form of echolocation. Sometimes I sing to them and they give back with resonant mews and trills.

On a recent visit I was lucky enough to make physical contact with Maya and Inouk, the zoo's two older belugas. Each took my hand deep into its great mouth, by way of tasting and greeting me. Their soft pink tongues clasped my hand as their mellons (those dazzling white globelike fore-heads) subtly changed size and shape to scan me. It felt like they held the whole of me, their eyes on mine, as they gently took my hand. For days afterward, my hands tingled with warmth.

If I am to learn to live by water, what better teacher than a cetacean? If my brother and sisters are across a con-tinent from me, what better siblings than marine mammal kin? And if I am to metamorphose and try to transform my life into what as a child I only dreamed about, what better myths to live by than the Northwest Indian stories that tell of mysterious beings who metamorphose back and forth between animal and human kingdoms?

In my study, next to my wall of Dreamer, Niki, and Sara photos, between my whale mobile and drawing of a purple Grandmother Whale drawn by my niece Lauren, is a small black-and-white sketch of a woman swimming upstream, her body half salmon. Entwined in a surging school of salmon, she leaps and insists herself toward her homeland. Drawn by Caroline Orr, a local Colville tribal member and native painter, this inspired storytelling-in-art reminds me of all the other worlds that coexist alongside me: Salmon People's world, the sky world, this Sound's watery world. Maybe I can learn to swim between them. After all, Native Americans know that salmon are really people who live in underwater villages. It is these people who graciously change into fish each spring to give us our food.

Sometimes staring long enough at Orr's *Raven and Salmon Woman*, I feel silver scales glisten against my sides and am emboldened by the bravery of a being who seeks her birthplace while making the circle of death and renewal. Sometimes gazing into Dreamer's familiar eyes as she swims toward me, my room full of windows becomes an air-filled aquarium. And sometimes when Seattle is so low and gray, the misting rain so familiar, I know we human mammals here in the Northwest underwater villages are closely related to the Salmon People, the cetaceans, and all life in the sea.

Each of us has an animal totem, some are blessed with many. We can all summon our animals and they will come. They will be our brothers and sisters and live alongside us — even swim upstream to die with us so we will not be without a guide through that greatest change of worlds. Some myths say that our animals then enter other afterlife dimensions with us, and return to begin this Earth's journey all over again.

I know that claiming cetaceans as my kin is not just science, it's shrewd. Learning to be human and to know what I might become, I need all the help I can get.

The local storyteller Johnny Moses (Nuu Chah Nulth tribe) told this Salish tale at a gathering:

> Long ago the trees thought they were people.
> Long ago the mountains thought they were people.
> Long ago the animals thought they were people.
> Someday they will say . . .
> Long ago the human beings thought they were people.

Oil Spill Eulogy

Splashed across my kitchen table that week of March 1989 was the *Seattle Times* photo of a Prince William Sound loon, its fierce plumage slicked with the black slime of the largest oil spill ever to touch our shores. Defiant, doomed, the loon's red eye gazed up and held mine; or perhaps it was I who looked longingly back at the bird, my eyes blurring as I gave way to the shame, rage, and helplessness I've felt ever since the Exxon tanker *Valdez* came to its terrible rest in "safe" waters.

Nothing is safe now. Not the sea otters sprawled across dark shores; not the ducks and living birds frozen because their oil-soaked down can no longer insulate them; not the golden and bald eagles and peregrine falcons who are starving slowly, their intestines destroyed by their grim feast of blackened wildlife. Even ashore, the deer, dependent on sea kelp, may join the oil spill's cull.

The safety and strict containment promised in the oil industry's emergency plans now seem as hollow as the duck-and-cover bomb drills of the 1950s. And whether we're Alaskans who believed the oil industry's promises or a public with an eye to gasoline more than environmental costs, this catastrophe leaves us with a deep sense of betrayal.

That night as I sat down to supper with the blackened loon, I read the *Post-Intelligencer* report from Cordova's mental health center director: "People are feeling overwhelmed . . . and being frightened by their own tears," he said. In comparing the spill to the death of a child or spouse for these fishermen, whose boats bob uselessly along Cordova docks, the director continued, "They can't believe they're crying. They're unaccustomed to a man breaking out in tears."

At work that week, my lunch table was full of colleagues also expressing sorrow. "This spill makes me ashamed to be human," said someone. "Yes," added another, "I saw an Exxon truck on I-5 this morning and everyone who passed turned to glare at the driver." Then we all talked about what we could do — collecting donations for the bird hospitals, gathering more towels and Dawn detergent to help wash feathers.

Finally a psychologist friend of mine told me her clients kept coming in with vivid dreams of the oil spill. It affected people more deeply than Armenia's and Mexico City's earthquakes, perhaps because it was man-made. "I've never seen a disaster go so deep," she said. "People are sobbing in their dreams."

Perhaps because this spill is the first to seep so profoundly into our collective consciousness, that is also where we must meet it, face-to-face. It is right, of course, to deal with it as a political, business, and environmental issue. We must assign responsibility if no one will step forward and accept it. We can begin with a captain known to have alcohol problems at the helm of such a potentially hazardous tanker. There is a terrible symmetry to a man polluted by alcohol commanding a tanker that pollutes so beautiful a body of water. And there is some shadow side to the Alaskans, whose economy has so thrived with the pipeline, also being the ones to receive this ruination. Then there is the underside of American big business. It is small comfort (and even smaller-minded corporate ethics) to read that Exxon's farsighted insurance and liabilities are such that the company will easily weather any financial loss. Exxon absorbs so little, while it has left the sea to absorb so much.

That evening after my last supper with the loon, I stood on my backyard beach, calling the gulls. I fed them leftovers from a friend's wedding. I didn't consciously connect those gulls who hung in midair cawing over the baked Brie, toast, and cookies with their Alaskan brethren until I realized with a jolt that I was surprised to see them still flying, white, not blackened by oil. As I gazed up at the flapping cloud of birds poised over me, I knew what was missing in all this oil spill response — no one is saying "We're sorry" to the animals themselves.

Even though we humans don't eat oil, we depend on it for much of our way of life. On a symbolic, collective level, these animals who have died are being devoured by our own hunger. Deep in our psyches we know that we have not yet done our part in this predatory gift exchange. My father, who hunted and fed his family game all our childhood, taught us this: "You thank the animal for its sacrifice. It's like saying grace, except you also give thanks to the deer or elk because it lays down its life so you can keep living." Along with mourning these lost animals, we must also be grateful for their sacrifice in showing us this sad truth: we human animals are out of balance and out of control. A simple task or chore is in order.

Here in Seattle where so many of us live by water, we might take some time during these days to go quietly to our lakes, our own Sound, even the decks of our ferries, and say to the sea gulls, the great blue herons, the bald eagles, that we are sorry. And perhaps some of us can say something more: we will change our hearts. Only by changing our relationship with nature and her animals can we begin to accept responsibility and heal the savagery with which we have preyed upon the world. This oil spill is, after all, a logical end of a mandate we have tacitly given our businesses: do whatever you must to maintain our way of life.

What if, at our own private funerals, our partaking consciously in the animal and human gift exchange, we also changed our mandate to say that our survival depends upon all other forms of life? As Chief Sealth said, whatever we do to the web of life, we do to ourselves. And what if we also each proposed to give up just one act of personal pollution? Such rituals of mourning, gift exchange, personal responsibility, and individual action, might then begin to heal this horror.

In a recent Canadian book, *Renewal* (Theytus Books), the Cherokee woman Gua Gua La (Barbara Elene Smith) tells an old tale of a time when the peoples of the sea and land were one. In our arrogance and isolation we have believed ourselves alone. Seeing ourselves so separate we have claimed dominion over all beneath us.

But the sea is not beneath us. It is our first womb, where we came from and now must in our minds return if we are to establish a balance between land and sea. Gua Gua La's sea people's song to their land family has served as my funeral song. In it I hear the backyard gulls and great blue herons.

> Oh my Brother, What has happened?
> Once we were as one . . .
> But now you have changed
> You walk alone in the darkness of your own creation.
> You do not hear me when I call you
> My heart is sad.
> I weep for you my Brother
> And I sing of you in my prayers. . . .
> I know now that some day
> Your childrens' children will dream a great dream.
> They will sing to you at the dawn of a new day.
> They will sing of the Oneness of all life. . . .

Whatever we here on land can do in our private eulogies, our promises, and water ceremonies, might bring some balance. Our small memorials will not assign blame or forgiveness. They can simply say to these, our fellow sea creatures, that we are ashamed. We are so sorry; we are grateful for the gift; we see the sacrifice. And we will hold ourselves fast to learning this lesson that has cost them their lives.

Wind on the Water

*A*t any moment in this city surrounded by water, there are those who watch to see which way the wind blows. They are not weather forecasters or stockbrokers or city planners; they are simple sailors who wait upon an element as subtle and almighty as air.

One of my housemates was such a wind-watcher. I call her Lynettie Sue, in honor of a tobacco-spitting junior high biology teacher who once sounded out the syllables of her first and middle names, to her everlasting chagrin. "He was trying to pronounce it like French," Lynette said. "But it sounded like he was calling in the sows."

It would have been easier if that teacher had called Lynettie Sue what those of us who sail with her do — *Captain, My Captain*. Lynettie is barely five feet tall, small-boned, quick, and dark as a sparrow; but at the tiller she is a *petite sauvage*. Once I watched a crew of men, their faces pale and slack, as she laughingly laid the sails — storm jib and all — horizontal to the water, took on several waves, then deftly righted us, tacked, and finally turned toward home. "This is the way we swab the decks on this boat," she explained. Leaning back to take the till with her foot, she called out, "Beer for the skipper." And the drenched crew, heeled over and hanging onto the rails, passed the Little King ale hand-over-hand to her.

Since childhood Lynettie has had a recurring dream: she is a princess on a royal barge sailing toward a New World. Perhaps she is Egyptian, or descended from those first Sea Kings said to have ruled over Atlantis. In her dream, the galley slaves listen only to her call. She's not just lounging about eating grapes or figs like most royalty; she's navigating, sighting on stars. Perhaps the salt water in Lynettie's family's blood runs to higher degrees of salinity than most. After all, her father, who is landlocked in the middle of Ohio, is building a forty-five-foot sailboat in his garage. By the time this book is afloat, Lynettie's father, with his nautical daughter and grandchildren in tow, will be sailing halfway round the world.

Sailing is in Lynettie Sue's blood; it is not in mine. Air is not my element so much as water. It had never occurred to me to put the two together until Lynettie moved into my house several years ago. I was living on Lake Washington. There I'd sit in my upstairs study, eyes resting as always on the water, when I'd hear a horrible racket at my door.

"Come out of there!" Lynettie Sue would call, banging

a buoy by way of entrance. "I need crew. And you need to play!"

"Can't you see I'm busy?" I'd begin.

"Can't you see the wind's up?" She'd shake her head as if I were deaf, dumb, and blind when right outside my window the madronas swayed, limpid swells on the lake calling. "How can you sit still when there's so much wind?" She'd study me a moment to make sure I wasn't dead, then she'd shake her head and sigh. "And I thought you were a sailor. . . ."

She'd drift out the door, but she knew I'd be in her wake like a sea gull. Soon we'd pick up another housemate or a neighbor or call a friend who sat watching out his or her office window — and we'd be at the dock. First we'd check in with Cap'n Jack, a co-owner of this old twenty-one-foot San Juan Two. It was the second San Juan ever made, but because of Lynettie's passion for shining teak and Tylex scrubbing compound, our boat barely looked as if it had seen a barnacle or brine.

"What say, Cap'n Jack?" Lynettie always greeted him as if he were the real skipper and she the second mate.

"I'd say . . ." he'd drawl and glance at the wind sock, narrowing his eyes. "I'd say you're in for a sweet sail."

Something about this little San Juan made Lynettie and Jack talk as if they were living in a sea chantey. A laconic drawl, a trancelike attentiveness to movement midair, which I couldn't see, a squinting, sweeping and continuous scan of the lake as if any moment a squall or ocean liner might trouble the waters — this was their sailors' manner I tried to imitate. But it took me two summers before I learned nautical talk, and I still haven't understood wind currents.

Once on board, Lynettie began her sail school. An ex-elementary school science teacher-turned-microbiologist, Lynettie on the water is infinitely patient. "Let's beat just for practice," she'd drill her crew in a series of precise tacks.

When she'd command us, "Ready about . . .?" we'd all brace ourselves, poised to tack, to hear her "Hard alee!"

Like a shot, I'd let out my sheets, line flashing through my gloved fingers like heat lightning. The white genoa would go slack, thwack toward the wind as I bent low in the boat then sprang to the other side to pull in the opposite jib.

"Haul in!" Lynettie would cry as I heaved with all my might. "Now trim those sails tight!" I'd brace my foot against the rail, arms aching as I heard the *chink-chink-chink* of the cleat catch and the jib go taut.

"Better . . ." Lynettie would nod. "But we lost wind."

Losing wind, running before the wind, close-winded, windward — the secret of sailing is wind. But it was always invisible and who knew where it was coming from or where it would go? When it was my turn to take the helm, Lynettie told me to pretend I was a baby bird in flying school. I didn't have wings, I had sails. "Where is the wind?" she'd quiz me with her eagle eye.

I'd gaze up at the telltales, those two strings of red yarn streaming from our shrouds. "Uh, from the left," I'd guess, as dyslexic about directions as I am about *p*'s and *b*'s.

"No, no . . ." Lynettie would softly correct me. "That's south. Now find your point of sail." I'd fix on Mount Rainier's benevolent white dome or even the Boeing Renton plant blooming up like a concrete mushroom cloud behind our backs. "See that ripple in the sail?" Lynettie would ask. "That means you're wasting wind. Turn into it . . . that's right, good!"

Then she'd fill my mind so full of scientific diagrams, diagonals, sailing facts that I'd lose track and turn straight into the wind, letting the sails billow and shudder like a shipwreck.

"You're luffing . . . you're luffing. . . ." Lynettie would cry mournfully, as if this lifeless, loud flopping of our sails really did signal we were dead in the water. Surely it was a sailing sin.

The teacher in Lynettie Sue was cast aside for the sailor. She'd take the tiller away from me, turn into the wind — which always told her exactly where it was — and we'd be heeled over before I found my seat.

"All right," Lynettie conceded at last after two summers of noting my mental block against the nautical science of knowing which way the wind blows. "Close your eyes. Do you feel the wind?" I did. "Now, follow it." I opened my eyes to see my hands had made subtle adjustments on the tiller and we were flying — running with the wind. It was a moment of astonishing clarity, like that first time I balanced on a bicycle. *Wind is like water*, I realized. If I can feel it, I can flow with it. Sailing had more to do with sensing than seeing.

I had a sudden memory. I was a child in North Carolina's Blue Ridge Mountains. Up there folks were so used to the wind, they learned to walk with it always at their backs, to lean into it to right themselves, to let its warm chinooks carry them when they were weary. My summer in the Blue Ridge, I instinctively learned the ways of the wind. So well did I learn that one day when the wind stopped howling down through those smoky-blue hollers, I fell flat on my face. I had forgotten to compensate for the fact that there was no warm air to buoy me.

"We'll just have to call you a psychic sailor," Lynettie smiled when she saw that sailing for me was not a conscious act. "We just won't tell anybody I taught you to sail with your eyes closed like you got some kind of sonar."

From that day on I was a sure sailor. If someone asks me how a sail works, I still cannot say — but I can sail. It's not in my blood, but it's in my hands and body. What is also inside me now is a reverence for wind.

It's an ancient and worldwide awe. There's a reason why in Italy when the hot, dry, and dusty sirocco wind blows, crimes committed under its sway are given certain consideration. There's a reason why women in those South Dakota, Kansas, and Nebraska flatlands who huddled inside earthen soddies against windstorms lost their minds to prairie madness. And there's a reason why the despair of this country's Great Depression found its metaphor in wind-struck wastelands of the Dust Bowl.

Wind, like water, is a Changer; and when the two elements come together, it is creation of the most alchemical kind. In many myths, Wind and Water are primal mates whose lovemaking spawns the Earth. A Finnish legend tells it this way: one day the virgin daughter of the King of the Air fell or leaped down through the sky into the vast oceans below. She wafted, this giantess, rocked back and forth between Wind and Water, who both desired her. Wind stroked her long hair, Water tenderly held her aloft, ululating while Wind whispered. Between them she was touched in all her secret places until her body opened to them both. Then the tempest as they took her — Wind was wanton and Water wild. When their storm was over, the giantess was no longer a virgin — she was the Water-Mother who drifted for eons, her belly wide and swollen, until she gave birth at last to all life.

There were days during my first two sailing summers when I imagined myself like that first woman, tossed about between wind and water. But it wasn't until my third summer crewing for Captain, My Captain on Lake Washington that I really understood what sailors call *heavy air*. There was a violent series of strange summer squalls, which eerily echoed our personal storms.

While June and July skies sent rain slanting down against our storm jib, when we wore slickers over our swimsuits, when crew were begging off their duties with a glance at the glowering low-slung gray skies — our little San Juan suffered a series of mishaps. I should have seen the boat was trying to warn me off the foul weather and rocks awaiting me on shore.

I should have recognized signs the way a sailor anticipates a change in the wind. One Saturday after a wet, slick sail, Lynettie Sue was maneuvering our boat toward the dock — "parking it," she called our perfectly choreographed dockside dance. At her shout, I'd lower the jib and we'd let air ease us toward shore. As Lynettie turned us alongside the dock, I'd lean out with a long pole and like threading a needle, I'd snag the bowline. We prided ourselves on this parking job. But that afternoon, the jib got jammed, I missed the bowline, and Lynettie's last-minute pirouette was skewed by a sudden gust of wind. Full force, we slipped sideways and slammed into the dock, one sheet catching on a post that lashed us fast while waves drove us time and again against the dock.

Like a small toy, our boat was wind-and-wave whipped. With each long scratch against wood and fiberglass, we all winced. Our heroic acrobatics only got me a smashed knee, and Lynettie's legs, arms, and hands were badly rope-burned. By the time we hobbled off the boat we were too stunned to mourn the gashes that scarred our boat.

That was the beginning. What happened then was that Lynettie Sue endured a series of lost love affairs that left her feeling like a castaway. On my part, the novel I had been so diligently writing when Lynettie first crashed into my study was rejected. Not with generic naysaying, but with three- and four-page single-spaced denials, all the more bewildering because each editor believed the book belonged to a different character.

Buffeted by so many strong opinions, I felt like a small boat besieged by bad air — that's when the wind blows from so many changing directions at once, that smooth sailing disintegrates into a series of short, staccato tacks. Those summer evenings, I'd come home with dread in my heart. In the driveway, I'd draw a deep breath, which I held until I rounded the garage. There I stopped, staring at the front porch. Twice a month, from June until August, I'd see a small brown mailbag slung up on the welcome mat. My book, my seven-years-in-my-study book, back, unwanted.

Once when I was living at the tip of Manhattan, I walked through Inwood Park and saw a bloated body floating down the Hudson River — bowler hat and all. When I'd called the New York City police about what I'd seen in the river, they said, "Happens every day, lady. But if it gives you any consolation, nobody saw this body but you."

And no one would see this book but me, I would think, and gather up the mailbag in my arms like a drowned thing.

One day that summer, I came home to find Lynettie Sue sitting on the front porch alongside another mailbag. "No wind," she said glumly. "We might as well be in irons."

I knew from her face she had ended another love affair. "Well, we're in the same boat, Captain, My Captain." I sat down beside her, holding the heavy manuscript on my lap. Hesitantly I suggested, "I guess we could motor. . . ."

Under usual circumstances and weather conditions, this would have been sacrilege to Lynettie Sue. But she turned to me with the mixture of giddiness and despair of any blasphemer. She said, "Who cares about the wind anyway? We don't have to wait around for anybody or anything!"

At the dock Cap'n Jack helped us heft the outboard motor. The day was flat and hot. He did not chide us for resorting to something man-made like this noisy machine that chugged us along when we should have glided, that choked and sputtered when we should have been soaring with our heads thrown back, skimming the water.

"Motorboats. . . ." Lynettie said it like a curse. "Motorboats and men . . ." She scowled and broke out the Jack Daniels. "They just trouble the waters."

"Publishers . . ." I nodded. "They're afraid to trouble the waters."

When we got to the middle of the lake, Lynettie Sue snapped off the motor. The lake was so still and forlorn; its wind-lover had left, too. A few motorboats cruised by, but avoided us as if sensing a ghost ship. Between the two of us, we ate a bag of Doritos and canned bean dip, Kasseri cheese and crackers, apples and grapes, and two chocolate chip cookies as big as our faces. We drank that quart of bourbon chased with mineral water. On land I cannot hold my liquor: put me in the air or on water, and one drink does me in. My memory of that afternoon is dim. I do know that Lynettie Sue and I sang the entire musical scores of *South Pacific* (with an emphasis on "I'm Going to Wash That Man Right Out of My Hair") and *West Side Story* (with special sneers on "A Boy like That"). We finished with a slurred and elegiac

Under the bridges
Over the foam
Wind on the water
Carry me home . . .

And with that we fired up the outboard to motor home. It gasped and stopped. "Reminds me of someone I know. . . ." Lynettie said savagely. She tried to crank it up again. Nothing.

We never did find out what ailed our outboard — maybe it had inhaled too many whiskey fumes. My mind was hazy from drink, but I do remember Lynettie making a solemn oath.

"We won't ask for a tow in," she swore. "We'll wait on the wind — it's wrong to be here without it. It's like . . . like being unfaithful."

Like her boyfriend. Like my publisher.

We were becalmed for six hours. We were thirsty with

water all around; we were alone even when other boats motored by; we were invisible as the wind. Maybe it was my besotted mind, but I sometimes thought the wind was still with us, even though we couldn't feel it.

Through parched lips I told Lynettie my suspicion.

"Sure," she said. "Wind's still here. It's air, isn't it?" She shrugged and sighed, downcast. "It just doesn't want to have anything to do with us."

"Lynettie," I heard myself say rather drunkenly. My skin felt warped from the sun. "Do you remember last summer when I broke up?"

Lynettie spat in the water and shook her head. "I told you you'd get your heart broken."

"I know, I know . . ." I scooted over and took the till. "Well, I think we should call this a memorial sail . . . you know, like those funeral barges."

Lynettie brightened. "We could have ourselves a wake," she said.

We did. We flatly floated there, dead in the water, and each told the other the story of every love affair gone awry. She told me about her lover with one leg who cut her off, too, and the golden-haired boy who slashed their waterbed by way of saying good-bye. I told her about the man who left me with the words "As long as I can find myself a dutiful wife, God forgive me, I will. I can't have two typewriters in one house."

We talked and told tales that might have put mariners to shame. We talked up a storm — we two women becalmed without motor or mates. And sure enough came a slight whisper over the dull horizon. Then a darkening in the water.

"Look! Look!" Lynettie pointed to a distant point in the lake.

I saw nothing, not even a ripple. But she had her sailor's face again. Squinting, she cocked her head to listen.

At last she cried, "Wind off the starboard side. Tack! Take it!"

I saw nothing, I felt nothing. But I took the tiller and closed my eyes. It was like imagining a movement and following it with my mind. Sweat dribbled between my breasts and my bikini top bound me too tight. But suddenly I felt a cool finger tracing my throat — lightly, lovingly.

I opened my eyes. Above me the mainsail swelled gently, as if blown by the faintest breath of some holy being who lived far away. And then I felt it full on my face.

"Jibe ho!" Lynettie called. We both ducked as the main swept across the boat. "Now!" Lynettie exulted. "Take us home!"

A gust filled the genoa, the till trembled in my hand. Tacking toward this wind, I found my point of sail. At last I saw the wind on the water — ripples, then waves splashing over our salty, sweaty skins. Fresh water and new wind. We were off.

At the helm I heeled us over until we all lay back, gazing at a Mount Rainier aglow in swirling azure and pastel pink skirts. We sailed like that, exuberant, skimming the lake, until we sighted on the dock, saw the silhouette of Cap'n Jack waiting for us. Above him the red wind sock flew full force. Only then did I relinquish the helm to do my dock chores. We parked, as we had in the past, perfectly. And though both my captain and I weaved on our land legs, Cap'n Jack didn't say a word about our reek of whiskey.

"Well, I was about to come after you in the Whaler," he said. "But then I remembered most women are half homing pigeon."

"Someday we women will run off to sea, Cap'n Jack!" Lynettie laughed.

"Men are always waiting on women," he drawled.

"Not you, Jack," Lynettie said and embraced the older man briefly. "We'll take you with us!"

Wrapping the mainsail, Jack at last asked, "Well, what happened to you all day out there?"

"We sailed off the edge of the world," I told him.

"Funny thing about being so long on water . . ." Lynettie Sue said softly, "We got heeled."

That next summer my book was accepted and we had another memorial sail — this time with no outboard motor, only the lightest, swiftest air imaginable. On a week-long sailing trip in British Columbia, Lynettie Sue met a man who, in her words, "filled my sails." They're married now and living in Vancouver. But they come down summer weekends and we all meet on the dock for another season of sailing. Lynettie's husband calls her captain, too, and he is a good mate. Recently, they bought Cap'n Jack's Boston Whaler as a wedding anniversary present for themselves and their new house. They live in a float home so that they will always be sailing.

I live now on this wind-shushed shore of Puget Sound. Drafts and currents breathe through my study, which is made of windows, all of which worship this blustery marriage of wind and water. I'm writing another book and hoping it will sail. I consider myself now a sailor and writer — for both ways of life require that we stare long hours out our windows, watching what is unseen.

Believing the Bond

Midnight, full early June moon. I'm walking my backyard beach, bare feet slapping wet sand, then sucked in if I stand still—which I do to stare at this amazing moon. It woke me with its strong, insistent light, shining into the living room where I'd fallen asleep on the couch, cat draped around my head like a Davy Crockett hat.

I walk carefully on the dark beach to avoid cutting my feet on black mussels and barnacles and because my housemate and her husband are sleeping on the seawall. It is low tide, a minus 3.2 feet that had us scuttling all day to dig butter clams and a geoduck I threw back because as it spat its last grit in the bucket, I swear I heard the misshapen creature screaming for the sea.

We tell ourselves we are in heaven in this house, here our first summer on Puget Sound. As I walk the midnight beach in bright moonlight, I feel full and illumined as the moon itself. This is a night I will remember on my deathbed — the stillness, the shining, the water shushing.

Now it happens: a shift in the sand, a trembling in my feet that shudders up my legs and swirls in my belly like radiant heat. Then deep, rhythmic jolts of electricity as if I'm plugged into a hidden socket by a vibrating cord. Surge after surge of energy flows up through my body moving with the waves. *Struck by lightning!* I think, though there is no storm, no flash, except the sudden splash of phosphorous gleaming against me.

Hands on the Earth, that's what I was taught. When lightning strikes, bend double and ground yourself by planting hands flat on the Earth; then you're like a small human rainbow, electricity running from Earth through your body and back to Earth. Because you connect, you survive the shock.

I plunge my hands into the Sound, palms flat against the grainy, falling-away sand. Feet and hands on the sea bottom, I let the pulses pass through me. I've never felt anything like this before. And yet it is familiar. I am afraid it will overwhelm me and my body will be blown out like an insignificant throwaway fuse by this electricity.

Bent double, my whole body shakes with the force of the pulse and at last I recognize its source. Not the moon, not the sky, not an electric storm, but the Earth herself. I close my eyes and unclench my jaw, letting the blood rush to my head. My hands are numb from the cold Sound, but they tingle with electricity. My teeth begin chattering and

I say silently, *Please, it's enough!* and suddenly the current stops.

Unsteadily I stand, my palms indented with tiny seashells that adorn my skin like barnacles. Never again will this Earth feel the same to me. I remember now that I live on a vibrant, breathing being that is a greater body encompassing mine. Images come to mind from a childhood story — an Indian girl plays atop the round, hard shell of a tortoise that is really the whole world.

One day the turtle wakes up, pokes out its sleepy head, and spots the little girl on its big back. Will the great tortoise shake her off like a parasite or let the girl continue to live on top of it? They make a simple bargain: the girl can stay on the tortoise's sturdy, round back if the girl will always remember to sing to the turtle as it sleeps. If the singing stops, so does the symbiosis.

My father, himself part Seminole, French Canadian, and Cherokee Indian, told me this story when I was very small. He also taught me the earth was alive. While I understood this with my mind, my body had forgotten the startling physical evidence of this living planet. Perhaps I forgot because it had been too long since I'd put my hands on the Earth and listened.

When I first learned to crawl, my hands moved along earth. I was born high in the Sierra Nevadas on a Forest Service ranger station. It was here in a small cabin I spent my early years. Wilderness was not someplace we visited, it was our home. When I crawled, I bellied my way along like a blind girl. What good was just seeing when you were crawling in spring grasses, eye-level with poppies, Indian paintbrush, and snakes? Better to use sound, smell, and touch to navigate the forest floor where pine needles and sharp-scented cones stuck in your palms, where flaming fall leaves crackled between your small fingers, and your hand perfectly fit into a mole's hideout or a squirrel's knothole home. Before I could name or objectify the forest, it lived in my hands. Before I had siblings or playmates, I chatted with stones and gossiping grasshoppers. My father called me "gopher," or "coon-baby," because of the mysterious dark rings circling my eyes for my first two years. They faded; the animal names did not.

My father's forestry work drew deeply on his Ozarkian farm-boy skills. He knew trees the way some people do finances or racehorses; he called them by name as if they were simply a tribe of very tall, silent neighbors who had their own ways. What he did not know was babies, and so he and my young, distracted mother raised me like the animals my father knew so well. If I cried, I was a coyote; if I sang, a mockingbird. And when I was the most inconsolable, my father suggested I hold on to a tree. Many of my earliest memories are of wrapping my arms as far as they would reach around a ponderosa pine, my face buried and criss-crossed with the rough imprint of its bark. The tree took the tears from me, transforming them into sweet-smelling sap I could chew like a child's cud.

I hardly knew I was human until we left the forest. And then, even though I later made my living in my father's footsteps as an environmental writer and editor and exchanged the wilderness of my first forest for the wildness of living by water, I forgot what the Earth feels like until this night so many summers later when I placed my hands back on the ground and remembered my first mother.

That summer night on Puget Sound, I considered another birth. It was a conscious connection, as if I, like that tortoise, simply woke up. And because it happened on Puget Sound's generous beach, I began to believe this body of water was another mother. My first love of the Sierra breezes through pines was transformed into the wash of waves, my original longing for the stalwart company of trees was changed into a fascination with the more fluid and secret life of the Sound.

After three years of living on this beach, I believed the bond was so strong I would always live here. I was wrong.

It began with the trees. Our landlord was rebuilding our deck. Slender, sturdy pine boards were stacked on the seawall and the drone of the chain saw was like a gigantic bee in the late August heat. There was pounding, then the lunging, singsong roar of the saw. Suddenly it was too quiet. Someone shouted, and just as I was running to the window I saw the tip of our small backyard cedar tilt crazily, then fall. In disbelief, I ran outside to watch our landlord, his face red with fury and the effort of lugging that heavy saw, stride from small tree to tree. In fifteen minutes, he leveled four — one western red cedar, which shaded my basement study and upstairs kitchen; another juniper, which protected our deck; and two small, lithe white firs by the fence, which allowed us to lie naked in summer sun, unseen.

The chain saw snapped off. Absolute silence. Our landlord stood swaying unsteadily, though there was no breeze. I was speechless, my head pounding. Perhaps if I closed my eyes, the trees would come back.

"They were out of control," Mr. Mulberry shouted. "The trees."

I will always remember the face of the landlord's son. He stood there astonished, staring at his father in fear and something else — submission. Had the tree massacre ended an argument between father and son? Did the father silence his son as he had the trees? Without another word, Mr. Mulberry threw the chain saw in his truck and rumbled away, leaving his son to stand quietly on the seawall, gazing down at some squiggles in the cement.

I knew what he was touching so thoughtfully with his boot — it was the clumsy, child's cursive of his own name carefully inscribed in the seawall cement: 1957, and the names of four children. Children's petroglyphs in liquid stone, the way the Southwest Anasazi pressed their painted hands against red rock: hands flat on the earth. *Remember me. I was here. Here's my mark in the earth.*

Then Mr. Mulberry's son came to the house, eyes lowered. He seemed very young, deeply ashamed. "I'm sorry," he said. "My father . . ." his voice faded off. "He's . . ."

"He's like the trees," my housemate said tersely. "He's out of control."

Then we went down to the water where Mr. Mulberry had thrown the junipers and firs in a fragrant heap on the beach. People who have lived for decades on this protected cove of Puget Sound habitually throw much of their lives into the water. Aside from the treated sewage and Saturday morning grass clippings, I've seen old clothes, paperback books, Clorox bottles, flowerpots, and once a busted-up cradle rocking by at high tide. Certainly some people assume the Sound is no more than a watery graveyard for the shipwreck refuse of our lives. We wondered that afternoon about giving our trees a backyard burial at sea.

As we hefted the trees onto our backs, trudging down the beach to where the low tide awaited her dead, I at last began to weep. Sea gulls cried, the wind picked up, and the cedar branches rustled as if repeating their own last rites. Suddenly I laid down my tree and ran to the house. I called my father in a way I'd never called on him for any other personal crisis.

Feelings come hard for my father — a man who gave his children long lectures on chestnut blight by way of warning us off certain dangerous lovers, a man who will present his daughter with a thirteen-pound wild Thanksgiving turkey he stalked himself instead of listening to why her marriage is breaking up. He is not a man his children call upon when in despair. He is good at giving money, what he calls "real world" advice, this being the world of career and politics. But like many men of his generation, my father balks at unmanly sobbing, what he calls "feminine fireworks," and tells us hunting stories as parables though we no longer live in the forest. His favorite censorship of any dinner table discussion of another family member's quirks or faults is to say, "Let's not have roasted relatives for supper."

So it was unusual to call my father for comfort that day. But it was fitting; it was not a lover or a friend who had died. It was the trees.

He heard the story in deep silence. Having worked in the Forest Service for almost forty years, ever since I was born, trees being cut is not seen as some environmental tragedy. But he, too, felt the unfairness, the waste of it. I thought my father might say something about it being the landlord's property and he could do what he wanted with every living thing on it — after all, wasn't that multiple use? Instead my story was received with the respectful silence I needed.

I took a deep breath and asked, "Can we plant the trees again?"

"Did he cut them at their roots?" my father asked, like a doctor diagnosing a wound.

Silence, when I said yes. "Well, Honey," he sighed, "it's awful hard to root fir and cedar again from branch, even though it's native, even if you do find good planting stock. Got another twenty . . . twenty-five years?"

I was quiet. "We're having a funeral," I said and felt suddenly small and silly, fearing he'd dismiss me like an overwrought child.

There was a pause, then he said softly, "Good."

"Shall we bury them in the Sound?"

"Water's another kind of earth," my father said. "Those trees might like to float for a while. It's hard work for centuries to send your roots far into the ground . . . sometimes down hundreds of feet. And you know what those roots are looking for, Honey? Water."

———

One morning the next summer, the exact weekend that was the anniversary of the tree cutting, Mr. Mulberry called to say he had sold our home. It was his right; we had a month-to-month lease. And though he had assured us he would never sell it as long as we stayed because he intended to retire there, though it was an investment he had held in trust for thirty years for his children to return to their childhood beach, he said his decision was based on the fact that "someone offered me more money than I ever dreamed of getting for that house. It's an offer," he said, "I can't afford to pass up."

"And your son," I said, too shocked to argue for myself. "I thought he . . ."

"No," Mr. Mulberry dismissed the subject. "He could never afford it. Not at this market's price."

"It began with those trees," a friend remarked when she heard we'd lost our house. "The minute Mr. Mulberry cut those trees, he not only stopped generations of junipers and firs, he cut himself off from his own children, his future." She was quiet, then added, "He's lost his family tree."

I thought of Mr. Mulberry's son tracing his own name in cement, I thought of me and my housemates like a little Anasazi tribe moving on, leaving only our handprints on beloved rocks.

As it turned out, the new owner of our house was also

devoted to Puget Sound. His first night walking the midnight beach at a minus 3.8 low tide, he stumbled upon a neighbor's illegal crab pot. Inside lay a mangled dark creature — a baby otter.

Just that morning, the new owner, with his wife and kids, had watched six otters playing off the breakwater. I'd never seen otters off our beach, I told him, only sea lions in their summer morning revelry. The new owner vowed to get to the bottom of the crab-pot poacher.

"And by the way," he added softly, "you're welcome here anytime. You all loved this beach. I know it's sad to be without it. . . ."

My last month in our beach house, a New York City friend visited and found me with my hands planted on the beach in my daily ritual of leave-taking.

She sighed and said, "It's *not* the same as losing someone you love. Puget Sound is not your mother abandoning or rejecting you. It's just a body of water that doesn't care one way or the other."

My friend meant to comfort. A childhood of being raised amid skyscrapers had convinced her my attachment to Puget Sound was at best a psychological projection and at worst an unnatural obsession. She wanted to free me from this watery way. Of course, the Sound was picture-pretty, a damn good investment as waterfront property, and it was comforting to hear waves at night. But there were other bodies of water, lakes and such, and certainly better houses.

"This old place looks like a 'Leave It to Beaver' house," she said.

I nodded and tried to make light of it. But the moment my friend was gone, I ran down to the beach and gave way to more despair than ever before.

Could she be right? I sat on the seawall and asked the water. The tide was low and sun glanced off the water in a long, brilliant fulcrum of light. I grew calm and felt the words deep within me: *Believe in the bond.* That was all that was asked of me. I would be allowed to stay on Puget Sound. Nothing else was revealed but that. And that was all I needed.

The next week, during research, I came upon a quote from Chief Sealth (Seattle): "How can you buy or sell the sky, the warmth of the land? The idea is strange to us. If we do not own the freshness of the air and sparkle of the water, how can you buy them? . . . This shining water that moves in the streams and the rivers is not just water but the blood of our ancestors. . . ." When I read this, I knew here was a voice that echoed my own love and loss. I decided then to go to Chief Sealth's grave and, after eight years of living on his land, at last pay my respects.

A short trip to Suquamish and the Port Madison Reservation, Chief Sealth's gravesite is not the embellished shrine we usually reserve for war heroes or statesmen. Yet he was both. During the bloody Indian War of 1855–56, which capped thirty years of warfare between settlers and Native Americans, Chief Sealth's heroism was in balancing his early embrace of the white settlers as friends with the needs of his own people. As recognized chief of many disparate Puget Sound tribes, Sealth refused to fight alongside others such as Leschi, whose lands were also being taken away in treaties drawn up by Governor Isaac Stevens to make way for the Northern Pacific Railway. Chief Sealth's statesmanship among his own federation of tribes as well as with white leaders is well known; his famous 1854 speech in response to Stevens's offer to buy his land rivals the eloquence of Abraham Lincoln.

Your dead cease to love you and the land of their nativity as soon as they pass the portals of the tomb and wander way beyond the stars. They are soon forgotten and never return. Our dead never forget the beautiful world that gave them being. They still love its verdant valleys, its murmuring rivers, its magnificent mountains, sequestered vales, and verdant lined lakes and bays, and ever yearn in tender, fond affection over the lonely-hearted living, and often return to visit, guide, console, and comfort them.

Such loyalty to the land and sea transcends even death, certainly it would survive a real estate boom. That's what I thought as I stood by Chief Sealth's grave that late summer afternoon. I realized then that this visit was my way of commemorating last summer's sacrifice of the trees, my way of consoling myself over losing our beach, and finally my way of asking guidance for the transition to come. As I had been taught, I brought offerings to this great chief's grave: blue corn from his Indian relatives who still cling to the cliffs of Arizona's Hopi mesas; a feather from a backyard sea gull, since this bird is said to be Sealth's totemic animal; braided sweet grass; a conch shell from our beach; and sweet peas picked from our garden.

Sealth's grave is dignified, but modest. Above a stone are raised two carved cedar canoes. Stately, they seem to float mid-sky. It is a simple graveyard, with many nameless graves like square granite footsteps left in the grass. And next to this cemetery is a small white wooden church. Chief Sealth's grave is on a mild bluff gazing out over Puget Sound to a skyscape view of Seattle. How fitting that this city took upon itself Chief Sealth's name.

Described in 1833 by a young Scotsman, Dr. William Fraser Tolmie, Chief Sealth was "a brawny Soquamish with a Roman countenance & black curly hair, the handsomest Indian I have seen." In the bronze sculpture at the corner of Yesler and First Avenue, on the square fronting the historic Pioneer Building, Chief Sealth's statue shows this strong beauty. But in another photo from the Suquamish Museum's book of their exhibit "Through the Eye of Chief

Seattle," quite another man peers out through half-closed eyes. Perhaps the old-fashioned shutter caught him mid-blink; more likely his eyes are half-lidded by age or emotion. Sorrow? Despair? Is he dreaming as he sits, hands clasped over a woven basket that looks like a great seashell on his lap? His hair, parted in the middle, falls in gray-black tangles to his shoulders.

He is a small man with strong mouth and brow set in benediction or thought. Is he remembering that morning as a child when he met Captain Vancouver, his first white man? Vancouver traded amicably with Sealth's tribe on Bainbridge Island, but later in his journal Vancouver dismissed them as ignorant. Sealth was not to see another white man until he was in his mid-sixties. That was when he helped a straggling band of whites survive the cold fall of 1851. John Low, Lee Terry, and David Denny landed at Alki Point. Low, a New Yorker, had hired Denny and Terry to build a cabin for his family-to-come. One glance at the rich forest lands bordering Puget Sound, the fertile valleys, the opportunity for sea trade and commerce, and Denny (a native of farm-rich Illinois) dispatched a note to his brother: "We have examined the valley of the Duwamish River and find it fine country. There is plenty of room for one thousand settlers. Come at once."

Denny, awaiting his kinfolk, accidently hacked away at his own leg while chopping firewood. He was nursed by Chief Sealth and his tribe, many of whom moved from Bainbridge to Alki, as they often did to set up fishing villages or tribal potlatches. According to the historian David Buerge, those first settlers Denny summoned to Alki stood "before the rain-soaked cabin, six men, six women, and twelve children huddled. At the sight of the roofless cabin, the crowds of half-naked native people, and the somber, dripping wilderness, the women broke down and wept." It was Chief Sealth's tribe that offered succor, staying through that difficult winter, teaching the new settlers the ways of this wild, watery new land. Some historians speculate that those first Alki Point settlers would not have survived that winter without Chief Sealth's people.

It only took four years from that fateful first settlement to the banishment in 1855 of Chief Sealth's tribe to their small Port Madison Reservation. Four years to forget such Native kindness. Four years to decide that this land was not to be shared, only divided and conquered. Is it any wonder that in that late photo Chief Sealth's eyes are almost closed?

As I stood beside Chief Sealth's grave in that shadowed sunlight through the great oak growing above his tombstone, I asked for teachings as those first white settlers did another century ago. I felt that "fond affection over the lonely-hearted living" was lent me. As I left, I put my hands on the mounded earth and listened. There is no translation for what I heard. It was simply a knowing that no matter where I am moved, blown about, no matter how far away I wander, Puget Sound is my true homeland. I will always return to her the way Sealth's dead never forget the beautiful world that first gave them being.

———————

A week after my visit to Chief Sealth's grave, I began looking for a new home. The first house I looked at was the first house we took. I gaze out now through a wall of rattling, double-pane windows that front Puget Sound. This beach cottage is so exposed, sitting right out on Alki Point the way it does. Sea gulls and Canada geese are swept sideways past my windows on forty-five-mile-per-hour winds. Waves thunder over the seawall, sending spray up to my attic study.

"Just wait until the winter storms," says the landlady. "You'll scrape seaweed off your windows."

I look at the native red cedar shakes, long known by Indians to shed rain and to give the best shelter on this windswept, watery shore. "The first cabin ever built here on Alki by white settlers had these same cedar shakes," I tell her.

The landlady nods. "There's a marker to those settlers a few blocks from here, says 'Birthplace of Seattle.'"

We both gaze at the vast, elemental stretch of Sound —from Vashon to Bainbridge and far beyond to Elliott Bay. "What's that island?" I ask.

It faces us, a hump of shadowy fir trees like a great moss-backed whale spy-hopping in the middle of the Sound.

"That's Blake Island," she says. "The Indians still have mighty good salmon feasts there."

"Yes," I murmur and smile. "Birthplace of Chief Seattle."

At night the wind shakes my attic room and my bed shudders as if am sleeping on a train. Sometimes the salt brine on our windows is so dense we have to scrape it away like frost; and just last storm our neighbors lost their satellite dish and front windows. Driftwood accumulates in our yard as we face the sea.

I sit in my upstairs study, which rattles so loudly in the wind, it's like working in a weather station. At my type-writer I often wear a hat, down vest, and gloves with the fingertips cut out; I imagine I'm reporting from the eye of a hurricane. I watch people seek out this Sound. Joggers, lovers, divers, secretaries, and construction workers on their lunch hours, old couples walking hand-in-hand, bent double against the wind. Once it blew a bundled-up passerby flat on his back like a beetle.

Storms draw the biggest crowds. Sometimes hats go flying and spinning like kites or crows. My neighbors patrol the beach for driftwood. One hasn't bought a winter cord of wood in years. We've sighted schools of dolphins and pods of orcas. I don't look for Chief Sealth; he and his people are everywhere here. If the Sound is my spiritual mother, then Chief Sealth is surely my spiritual father. My own far-off father would understand; in fact, he's thinking of a fishing trip out to the Sound.

Walking this wild, seaweed-strewn beach, I sometimes remember that my blood is very similar in composition to seawater. I am, after all, evolved from an ancestral amoeba only recently emerged from primal slime. According to geologic time, I am a relative newcomer. Who knows how long my kind will last?

Chief Sealth said, "There is no death, only a change of worlds." We will all survive in some shape or form, maybe hardly human, maybe barely recognizable, maybe never dreamed of. Maybe we'll haunt the land we love as rock or tree or sleeping turtle. Maybe we'll survive whatever shock, whatever change of worlds because we connect by touching the Earth — because we remember our long-ago treaty with that turtle and keep on singing.

Watching for Whales in Winter

"*C*all the whales with your mind," suggested my friend Betsy Dayton. It was mid-February and we were sitting on the rocks at San Juan Island's Lime Kiln Point State Park, where whales come close to scratch barnacles off their sensitive skin. "Use your third eye," she laughed, "your sixth sense."

My winter whale-watching sojourn had begun only a few months earlier, when I'd sat with Betsy, an acupressure practitioner who lives in Friday Harbor, in a cozy Seattle bar drinking hot toddies by a fire. I told her of a dream I'd had of a wind-bent giant bonsai tree on top of a blond bluff overlooking a wide channel of dark violet water where two great rivers met in the mist.

"It's salt water," I explained, "but in rivers, not the sea. The dream is like being inside those painted Chinese silk scrolls." As I spoke, I doodled the dream image on my napkin, crumpling the picture in frustration. "Have you seen any tree like this in your travels?"

Betsy smiled and sipped her hot toddy. "Come visit Friday Harbor," was all she said. "We'll watch for whales."

"In winter?" I demanded. "We won't see any whales in this weather."

"Never know." She smoothed out my napkin sketch. "A few whale pods winter in the San Juans while the other families migrate south. Come watch with me."

And so I found myself in February in Friday Harbor on a rain-swept Saturday, crouched among cold rocks, squinting against wind and mist. Nothing was on the horizon except fog-dimmed shapes and the suggestion of Victoria, B.C.

"Some scientists theorize from prehistoric skeletons that whales were once land mammals who returned to the ocean," Betsy commented.

"Maybe whales knew something dinosaurs and our ancestors didn't," I said. "Don't whales have the biggest brains on Earth?"

All afternoon we stared into the vast wash of gray, as if it were a dense Northwest curtain about to part and reveal black dorsal fins arcing above slick black slaps of orcas at play. We sat there on the jutting rocks, eyes half-lidded, skin swathed in goose down, ears cupped by sheepskin muffs. We tasted sea spray and smelled the pungent winter beach fragrance of seaweed. All senses were blurred and fluid, like being under water.

Was this what it felt like to be a whale? Living a long, liquid life inside a vast ocean, tons of blubber protecting the hot mammalian blood, swimming easily to unfathomable depths in a manner that we land mammals might call unconscious simply because humans black out at the bottom of the ocean? Every breath a whale takes in its life is a conscious act. Whales don't have the autonomic nervous system that breathes for man without his will. Humans can last years in a coma; if a whale loses consciousness, it will drown. What does this say about the enlightenment of our fellow mammals? Who is conscious and who is dreaming on small shores while whales claim the watery seven-tenths of this Earth's territory?

"Listen," Betsy said suddenly, breaking my meditation. "Hear that?"

A deep blast of air and sonorous inspiration in the mists, several times, then silence. "What big, beautiful breathing," I murmured, feeling strangely comforted.

"Now," Betsy said softly, "call them with your heart."

Then she told me about how whales scan inside one another's bodies, especially the organs, to determine the emotional state of a cow or bull or newborn calf; how the cetacean brain has a highly developed paralimbic lobe (missing in man) that works like a computer to combine all separate senses into one multifaceted perception. Whales can use their perceptions — what we might imagine as a three-dimensional hologram — to "see" another pod several hundred miles away through echolocation (like our sonar), to "hear" and translate the song-histories among humpback whales or chatty dolphins. The whales' skill at sensing what is unseen is so refined that it leads some scientists to wonder if our fifty-million-year-old cetacean relatives haven't fully developed what in humans would be called extrasensory perception or a sixth sense.

What humans perceive as psychic or *a priori* knowledge might simply be a part of a whale's way of life. As we waited on those rocks, I realized that everything I've truly intuited has always come to me after surrender to an attitude of openness, receptivity, deep feeling — an attention so still it is like prayer or meditation. Willfulness, force, and goals have nothing to do with intuitive knowledge. Perhaps our human sixth sense is a synthesis of all other senses (as well as the mind and heart), which trusts the world as if it were at one with us, our greater body.

"By the way," Betsy said, as we at last stood up — it was evening and the lights of Victoria sparkled like a human constellation eye-level with us — "two rivers meet right here."

I glanced at her, the nape of my neck prickling with goose bumps. "What?" I said. "This is salt water, isn't it? The Sound?"

She laughed. "Salt water, yes, but running in rivers. The Haro Strait meets the Strait of Juan de Fuca right here. Come on. Let's watch the sunset from those cliffs."

All senses heightened, I scrambled behind Betsy. Suddenly I felt a rising elation.

"My friend tells me there is another river-meeting like this in the Amazon," Betsy called. "The natives believe it is holy and call the place 'The Wedding of the Waters.'"

But I was no longer following her. My body moved without my will, as I ran ahead up the bluff. My heart pounded, then opened like a great eye as I scanned the cliff and then saw my Chinese tree — this wind-bent bonsai fir, many gnarled limbs reaching for earth.

"It's here!" I shouted. "It's always been here."

"Yes," Betsy said as she climbed up to sit with me under my dream tree. "Like the whales."

Very still, watching the water, we sat overlooking this place where two great rivers marry and where whales winter. For me the ceremony had begun, this marriage of true minds: one human and small, and one the biggest mind on Earth.

Now when I go whale-watching under my dream tree in Friday Harbor, I believe that the whales in those cold, deep waters are also watching us. They wait for us to wake up, to follow our dreams and sixth senses, to meet them as kin. Perhaps they'll help us forgive ourselves and so find family reunion. After all, whales are the most great-hearted mammals on Earth.

Wild, for All the World to See

*I*f in Frost's New England "something there is that doesn't love a wall," then there is also something in the Northwest that doesn't love a porch. It's not that we don't relish living outside with our waterfronts, our casually stunning views of Mount Rainier, our tenderly tended shade or rose gardens. For some reason, we in the Northwest prefer those spacious cedar, cottonwood, and pine decks, which gaze meditatively out on nothing more public than our own backyards.

Why do we shy away from porches, even those of us raised in the sweeping veranda traditions of the South, New England, and along the Midwest's loquacious, elm-lined streets? Why didn't the architects of the early Washington settlers reestablish posthaste the farmhouse porch, upon whose creaking, much-memorized boards they could while away a warm evening or gracious Sunday afternoon?

I suspect the answer lies in the character of a porch itself. Porch is persona — that prominent declaration of a private life. As persona then, a porch by definition can be a facade, masking the true character. In this pretense there is often a gleeful make-believe akin to the stage and sports. Just as everyone knows without saying that TV wrestling is not exactly a real sport, so we also realize that what we reveal of ourselves on front porches is not the inner, unadorned family life. Porches are where we pretend perfection, what we would like the world to see and remember of us.

My grandmother used to dress up to sit on her porch — not the same as she'd dress for her Southern Baptist church pew, but a grand act of theater all the same. She took it upon herself to teach us porch etiquette by way of redeeming us each summer from our redneck (i.e., not town) other grandparents, whose porch was tastelessly placed in the backyard way too near the chicken coops and bordering a catfish-stocked lake my Ozarkian grandfather had bulldozed himself. Every evening Granddaddy fished for his supper. This showed an appalling lack of delicacy to my town grandmother, who suspected that only a savage would kick his boots up and gut fish rudely, right there on his porch.

"At least your grandfather built his porch in the back," my grandmother conceded. "Thank the Good Lord, he's not wild for all the wide world to see."

Being wild for all the world to see was a cardinal sin in small-town Southern society. Wild inside — let the china crash, the children slide down the laundry chute, the bed-springs sing — but oh, *not* on Grandmother's front porch.

Grandmother notwithstanding, wild things did go on set against the backdrop of that porch. I survived my first kiss on my grandmother's porch swing; it was a kiss more impressive for its acrobatic timing than romantic technique. And, sure enough, it was my cousin Douglas. His mother was my mother's next-door neighbor and his father and my father were brothers. So right there between the two proper houses we had a whole passel of summering cousins. The kiss, launched from the opposite end of the porch swing at the exact apogee of our flying arc, was perfectly planted at the instant I blew a Bazooka bubble. Blinded by the pink gum plastered even to our eyebrows, Douglas and I stumbled into Grandmother's kitchen, tart with its simmering apple butter scent, and she had to take witch hazel to our faces. As she scolded and scrubbed, I realized she had guessed what was behind our burst bubble.

"If you have to play like that," Grandmother dismissed us with a persnickity purse of her own lips, "do it in the root cellar, not on my porch."

So we learned, kisses were for the dark hide-and-seek and holding hands was the limit on the porch. This incident sparked what we cousins called Grandmother's "porch drills." Whenever polite company came calling, Grandmother rang a little bell that called us all from root cellars, from the next-door chicken coops where our cousins' grandmother amazed us with her afternoon wringing of chicken necks, and from the attic where my sisters still dallied with doll babies. We appeared breathless and stood at attention for our company.

"Children, take the steps," Grandmother would always say, as if this gracious gesture just occurred to her. "Let our guests swing." Actually we'd been well taught to sit on the steps in stoic stillness as if languishing on cement pews. Being outdoors, without even that wooden pew back to slump against, made us mutinous. What were we supposed to be reverent about or worship on the porch? Grown-ups chatted endlessly when we could be playing. Sometimes the adults were so achingly boring they fell into silence as if shot with ray guns set on stun. These adult lapses we explained as what happens when grown-ups are overcome by their own ordinariness. It never occurred to us they might be content or companionably silent as they smiled and nodded, staring at their front lawn sprinklers, the only conversation the crickets and creaking chains of the green-slatted porch swing. It was awful when adults took over our porch swing and swung it like a slow cradle instead of the amusement park ride it could be with the right daredevils.

Perhaps enduring these childhood usurpations of our porch explains my later teenage total uprising. At sixteen, I spent a mandatory summer at a Southern Baptist camp as a lowly waitress serving mostly Texans, who pulled up in buses with banners proclaiming "Wallace for President." I hated the camp, full of would-be missionary students. Some of the teenage boys were Jimmy Swaggarts waiting to hit the pulpit and the strips; and the girls actually planned prayer meetings on the porch of our Yucca lodge. Of course, believing prayer was a private affair, I never attended the evening porch vespers.

One night, however, I peeked out of my upstairs window and lowered a plastic baby doll we called Fat Albert — noose knotted around his pudgy neck — to startle the prayer meeting. When my accomplice and I sneaked down to see what horror we might have wreaked, we found our doll dangling unseen, while rockers and porch swings creaked crazily. The spirit of Cousin Douglas was there that night, because here was an entire porch of kissing couples. This was more like it! This was what a porch could be without grandmothers. I suppose I could say a little about the missionary position and the fact that in the South a woman's bosom is called by both sexes, her "front porch," but these details are better left to the imagination. Suffice it to say that this was the one detour from porch etiquette I ever observed — and I observed it faithfully.

I cannot now ever think of porches without remembering both my grandmother's strict decorum and that Baptist summer debauch. Watching those fervent prayer meetings was more romantic than the movies and more exciting than Southern lawn sprinklers.

And perhaps that's the whole point about porches: you are watched on a porch, no matter how wild or proper the behavior. Porches are society set smack dab in our own front yards. Because the porch persona courts an audience either consciously or unconsciously, there is really no place on a porch for private confession or the naked truth.

Not so on Northwest decks. Set sturdily in the back, shielded by shrubbery or trees, the Northwest deck is more often than not a sanctuary or haven. Hot tubs, secret sun worshiping, summer loafers clad in scanties or disheveled comfort clothes more suited for gardening than greeting guests, the Northwest deck offers itself for nothing more dress-up than a barbecue. Who would put on pearls and pumps for her deck? Who would meet a gentlemen caller or first date on a deck? Who would invite the neighbors in Sunday best to visit their deck? And most of all, who would furnish a deck with swing or rocker?

I suspect my grandmother would feel profoundly uneasy here on my deck that overlooks Puget Sound. Summers we live on the deck; many nights we sleep atop our inflatable double air mattresses called "entertainment centers" — a Costco must with forty-eight convenient cylindrical beer-can holders and plush plastic pillows. The bed can double as a seaworthy raft and at one of my birthday parties we had seven entertainment centers roped together, a flotilla of floaters on Puget Sound.

Last summer we did not have a weekend without a barbecue. At the end of a feed and in a savagery reminiscent of my country grandparents, we hurled our leftover hot dog buns and BBQ chips to the sea gulls, our most frequent deck guests. On this misty, private beach deck, I've made love — not furtively, fumbling in fear of what the neighbors will think or witness, but luxuriously, with only the ancient caw of the great blue heron, the shush of Puget Sound, sometimes the good-humored bark of a migrating seal. Dreaming on my deck, I hope the orcas and dolphins and occasional gray whale will blow a welcome going by. And one summer I stood on my deck and watched a vast sheen of phosphorous ripple, shining out like underwater spirits.

Still, with all these Northwest wonders, I often miss my grandmother's porch. Maybe it's the swing or the playacting or even the little *chink-chink* of sprinklers. Maybe it's that I don't believe in segregation of porch and deck. There seems only one solution for me, now that the Northwest is my chosen home. Why not, O architects of the Great Northwest, why not build another union of North and South? It may be an unholy marriage in design terms, but what a wonderfully mixed marriage: a front porch for receiving folks, for courtesy and personal reverence and courtship; and there hidden away in back, an expansive deck for wide-openness, for nature loving, for living wild for all the world not to see.

Where the Green River Meets the Amazon

*O*ther worlds often interrupt our most mundane routines. I was on my early-morning commute along the lush curves of the Green River Valley, which like my sleepy mind was still mist-shrouded. My car stereo played the soundtrack to *The Mission* — that heart-stirring choir of South American Indians singing their Jesuit-inspired "Te Deum Guarani," in the high jungle above Iguazú Falls.

With those massive waterfalls in mind, I slowed to make my sharp right turn onto a road that runs like a blacktop snake alongside the dark and verdant Green River. The disturbing music and foggy river road drew me further into darkness. Engulfed by gray, there were no other headlights, no movement, just the wide, winding stream like a shadow swirling at my side.

Sometimes it seems I am carried along the Green River backroads more by music than by my car. That morning, I barely noticed the dashboard clock inching toward my nine o'clock office deadline. Instead of to my gray steel desk, the mysterious music swept me back to the 1750s Amazon jungle where Brazil, Paraguay, and Argentina now meet at Iguazú Falls. A friend who flew over those thundering falls in a helicopter says of them, "The earth just caves in there — it's the edge of the world."

As I meandered along the Green River in the mist, I too was at the edge of the world. A world where Stone Age Indians still live naked in rain forests. For three centuries these Amazonian Indians have been enslaved, massacred, and invaded by Spanish, Portuguese, and other colonizers. As I listened to *The Mission*'s resonant choir and flutes, I remembered what haunted me most about the movie: a scene showing Portuguese and Spanish soldiers laying Indian children in the mud as if putting them down for a nap. But as the mothers wept wildly, they and their children were shot en masse. The last moment of the movie shows some Indian children who survived the mission massacre setting off in a dugout canoe, their only mother now the rain forest.

On the foggy Green River road, slats of smoky sunlight now alternated with dense floodplain foliage as I traveled my own jungle, with music my only guide. Just as the unearthly, piercing echoes of South American panpipes reverberated in the steel chamber of my car, a red-tailed hawk, wings wide as my windshield, scudded across the glass, its golden eye glancing mine.

I swerved, caught my breath, and barely missed both the hawk and a plunge into the Green River. Hands trembling on the wheel, I stopped the car. All was still. Then I started shivering, those tremors that begin in the bones. It was more than the ordinary fear triggered by a near-collision with that raptor's great wings; this fear was old. Perhaps the hawk, whom the Native Americans call "messenger," was simply reminding me of what I rarely allowed myself to face during these years of commuting in the Green River Valley: *Here is burial ground.*

Here as the panpipes sounded like lost souls, where the Green River's depths flow murky and slow; here where the music flooded my memory with Amazon Indians massacred in the mud as they sang and fought in far-off jungles — here is where the perpetrator of this country's longest series of unsolved murders takes his name and buries his bodies.

Those of us who work or live by the Green River have a keen though unspoken sense of just where we are. At midday, women and men from my office jog or walk, always in pairs. We do not discuss safety, it is only second nature. No one goes alone near the Green River, for all its well-maintained trails and pathways. One lunch hour, my co-workers returned with news of another buried body found that day. For a week after, most employees ate indoors, though the sunshine was brilliantly inviting. This summer someone tacked on the bulletin board an article about two men convicted of killing a young woman. According to the *Seattle Times* report, they hog-tied her, "beat her over the head, dragged her to the Green River . . . and left her to die." Then they "celebrated after the killing with beer and sausage." Were the teenage killers imitating their Green River murderer hero? Were they unconscious followers of a tradition even older than that of the Green River murderer? After this article appeared, I noted that my co-workers took to jogging in groups of four, some carrying heavy hand weights. Others began bicycling the backroads.

One afternoon, a colleague returned from his run to report a dead hawk on the side of the river road. I went alone to bury the bird, though I took my car. As I lay that still-fierce bird of prey in the mud, I wondered if this were the same messenger who'd met my eye the month before when I myself almost lost my life to the river. Even now I keep one of his elegant brown-and-beige-striped feathers on my dashboard.

———————

The Green River Valley has some of North America's most fertile farmland. Along the Green River banks, there are still single-family farms with U-Pick-M strawberry fields, tidy acres of corn, lettuce, beans, broccoli, bales of blond hay. Though today the valley's biggest bumper crop (besides bodies) is the concrete bunkers of corporate headquarters, my morning commute still takes me past marshy wetlands where herons stand poised on one leg and hawks swoop from rafter to rafter in abandoned dairy-farm sheds. From my office window, I contemplated canary grass, cattails, and purple wildflowers until bulldozers began their bass rumblings this last spring like some mechanical mating call. Some lunchtimes, I picked berries in nearby fields and talked with local farmers about the loss of their land to developers. They told me this valley was once a center for growing hops, and in the 1930s they celebrated a lettuce festival and another crop festivity called "Cornucopia." With World War II, many of the Japanese-American lettuce farmers were forced into relocation camps. In light of that wartime disinheritance, the current foreign Japanese investment in our real estate takes on a strange shadow.

But Japanese-American farmers were not the first to

be dispossessed here. Two hundred years ago, small Indian bands lived winters in large cedar longhouses along the riverbanks, fishing and hunting elk, deer, duck, and beaver. According to the Muckleshoot Indians, the rivers were so abundant with fish that at spawning time one could literally walk across the gleaming backs of salmon and pluck some up for supper. Five miles of the Muckleshoot Reservation, between Auburn and Enumclaw, is the only Green River basin land left to the once many and diverse Native tribes. Upriver, along the Black, Duwamish, and Cedar rivers, lived the Duwamish and Suquamish tribes, a union that gave birth on Blake Island to Chief Sealth in 1786.

During the Indian Wars, it was Chief Sealth's influence that kept coastal settlers and Green River or Puget Sound Indian tribes from escalating their attacks. For thirty years, the bloody skirmishes between whites and Indians raged. Early reports of these battles remind us that before the Green River was burial ground for a serial murderer's victims, it was graveyard for many Indians and pioneers. In the mid-nineteenth century, Chief Sealth predicted: "When the memory of my tribe shall have become a myth among the White Men, these shores will swarm with the invisible dead of my tribe, and when your children's children think themselves alone in the field, the store, the shop, upon the highway, or in the silence of the pathless woods, they will not be alone."

That summer morning when the hawk caught my eye, when the Green met the Amazon, the flow of these two rivers broke over their banks and into my heart, and I realized: I am not alone in this land. Bodies and souls, mud and massacres, history repeating itself is all around. Memories cling to these riverbanks like mist.

Primitive tribes have always said that the confluence of rivers is a special haunt of supernatural beings. So what happens when our rivers become burial grounds for massacred bodies, when they become waterways to carry away the carnage of their own forests, when they are developed, their flows willfully diverted, dammed? Does a corresponding sorrow then echo like otherworldly music inside our own ears?

Today, descendants of those Amazon Indians under siege in *The Mission* are dying at an alarming rate. Their deaths mirror the staggering rate of extinction among rain forest species. Since the 1970s' road construction into the Amazon and Brazil's blind plundering of her own rain forest resources, whole Indian villages have been destroyed by development, influenza, and deforestation. Imagine a Kreen-Akarore Indian of the Amazon basin who has never seen metal, coming face-to-face with a bulldozer and his first white man? Imagine the feel of godlike dynamite shaking the earth under one's bare feet. Imagine fleeing farther and farther into a fiery forest that disappears at a rate of fifty-four acres every minute? Last year, in our satellite photos, the Amazonian rain forests were seen from space as so many gasps of smoke.

The mighty Amazon River and her forests cover 2.7 million square miles, equal to 90 percent of our own country. Every time rain forest acres are burned, Amazonian tributaries dammed, and those breath-giving forests slashed, there are global consequences. World temperatures rise, signaling climatic changes; atmospheric carbon dioxide increases, contributing to the dangerous "greenhouse effect"; and millions of rain forest species, some long known for their healing powers, disappear.

Yet it is hypocritical for those of us in developing nations to decry deforestation of the Amazon rain forests. After all, this country's development of our own West simply predates the Amazonian exploitation by a century. What is perhaps different is that now we are becoming aware of what it is we lose when we slash and burn our old-growth forests, when we dispossess our native cultures, when we live only for ourselves and not our future generations.

In reading about our own region's Green River Indians, I came across a curious description of Puget Sound spiritual life: "Soul loss was a mysterious malady to which North Pacific Coast Indians were subject," writes Philip Drucker in *Cultures of the North Pacific Coast*. The soul's departure from the body "did not cause immediate death, but rather lassitude and wasting away, which was fatal if the errant soul were not recovered in time." It was up to the tribal shaman to bring back this straying soul, sometimes using a supernatural canoe, and so restore the body.

Do we lose our souls when we desecrate our land? And in those places like the Amazon or Green River basins where murders, mass burials, wars over development are all inexorably linked, does this brutal confluence also summon the supernatural by drawing down the darkness?

I lost some of my soul that morning along the Green River — and it is not restored. When all the rain forests have burned, when all the waters are slow and murky, stained dark with our dead, our refuse, will we at last recognize that we have conquered not the earth, but ourselves?

There is no more use in conquest; there are only new ways to connect. The first connection is very simple. It's a feeling that we are not alone, that history lives alongside our most ordinary comings and goings and that there are global as well as personal consequences from our actions, our dispossessions. If we harm, we will be harmed; this is not environmental karma, this is scientific fact. The natural world is our greater body. If we destroy the breathing rain forests, sooner or later we too will cease breathing. This bonding of or being with the Earth is not radical; we do it when we die by simply burying ourselves in the ground.

There have been no mass funerals down by the banks

of the Green River, just as there have been no mass funerals for all the species, tribes, and trees destroyed in the Amazon. Perhaps we need to lament our collective loss of soul. Perhaps that's the message the red-tailed hawk delivered to me that morning when I ran off my river road and into the Amazon jungle. Perhaps that hawk I buried and whose feathers I still carry on my car flies with me now as a daily reminder that I am never alone.

Daily, slowly, in small, seemingly insignificant rituals we might someday summon back our own and the Earth's lost souls — by holding one funeral for one unknown victim of the Green River murderer; paying one visit to Chief Sealth's grave to give thanks for the land the Indian tribes gave us; making one acknowledgment of the history, the people who died and are buried here in this land beneath us.

Late this summer I stopped commuting through the Green River Valley. There are many reasons I left my job there in that burial ground. But one of them was the realization I had that morning when two rivers met in my mind.

Nowadays when I do drive the Green River road, I do so with full consciousness of where I am. And now the music on my car stereo is different, too. In his deep-throated, sometimes bleak ballads of lost land and tribe, the singer Floyd Westerman, himself a Dakota (Sioux) Indian, descendant of Red Crow, sings a song in Spanish that might as well serve as a requiem for rain forests, tribal peoples, and those murdered alike. Sometime we might recognize the song as our own:

> *Mis hermanos,*
> My brothers,
> *La tierra es mi madre*
> The land is my mother.

Mis hermanas,
My sisters,
La tierra es tus madre
The land is your mother.

Mis hermanos,
My brothers,
Somos de la misma sangre
We are the same blood.

Mis hermanas,
My sisters,
Somos un solo rio.
We are one river.

Playing with Nature

Midway through my life, I found myself in a dark wood. I'd undertaken this journey into Schmitz Park, Seattle's last stand of virgin trees, to show a visiting Cuban friend, who had never seen snow, how these dark, ancient woods glow like a fairyland full of white drifts and bent cedar boughs.

Certainly I didn't expect fairies, but as we walked into the park, footsteps muffled by the dense whiteness, we heard a cry. From behind several trees leaped a band of medieval knights, their red crusader crosses blazing across white armor. On the snowy path, the knights bowed in deep courtesy to a gathering of black knights, a green-tailed dragon, various velveteen ladies-in-waiting, and a few elfin folk. Without a glance at us, the pageant passed by and we stepped aside.

"Maybe we are in a movie?" my friend suggested.

"But there are no cameras watching," I said. "Just us."

A large woman clad in black stepped from behind a two-hundred-year-old western red cedar. "Yes?" she asked.

"What *is* all this?" I said. "Who are you?"

"I am the Dark Force," she said softly and let out a low, lovely laugh. The laugh lowered, resonant and soulful as a cello — but I didn't think she was smiling behind that veil.

"Strange things happen when it snows," my friend whispered. Then she turned to the dark woman. "In Cuba it's carnival time. Is this your Northwest carnival celebration?"

The woman laughed lightly now. "We do this all the time. I usually play a gypsy alchemist. And those change-lings over there are cat people, the Kzin. These are their woods."

Indeed the woods were full of cat-faced, fur-clad people who crept stealthily along the creekbed. I was struck by their feline attentiveness to far-off cries, a snapping twig overhead, that bird swooping low. A Kzin scampered across the path, and seconds later we saw a tall man dressed in a richly embroidered scarlet cloak; a silver dagger dangled from his thick leather belt. He did not greet us. His eyes were riveted on some distant ridge where there was faint shouting.

"That is our Gamemaster," the Dark Lady explained in a hushed tone. Her voice was subdued — more from stealth, I guessed, than respect. "He has given me the task of persuading the people to kill their unicorn."

"Why?" we asked.

"Because that is what they love most."

She pressed a gold coin, stamped with the winged image of Pegasus, into my mittened palm. Then she turned and strode into the snowy woods. Pausing at the top of a hill, she made a mystic sign in the air, and I noticed she wore fluorescent orange gloves.

I would not recognize this Dark Force if I saw her again, for she hid her face from us. But I recognize the forces at work in those woods. It's what happens when we play with nature, using not our will but our imagination.

What we witnessed that day was an elaborately organized group-at-play called The Fantasy Alternative. Their motto is "Entertainment through Education," and at regular meetings they plot intricate games to play against natural settings. As adults they are simply continuing what we all began as children — playing in the woods.

My friend and I ambled along the white trail, pausing every so often to listen to the sounds of a pitched battle. Was it black against white knights? Was it the Kzin purging their woods? Were they the penitent cries of the people who had sacrificed their unicorn — a beautiful beast we never did see?

My Cuban friend, Flor, is a Los Angeles psychologist who has spent most of her career "counseling children and other artists," as she explains it. While we watched the Kzin dart in and out of the woods, she commented, "Our fantasy life needs to include nature." In her work with children of the *barrio* — where gangs offer a surrogate family and drama is acted out in the streets — Flor has taken these children into the wilderness to participate in simple rituals. Using the traditional Native American concept of the spiritual quest into nature to find one's own vision and *totem*, or animal guide, these streetwise children develop deeper wisdom.

"At first the kids are afraid," Flor said. "One tough little nine-year-old boy, Renaldo, asked me, 'Do they have Nintendo there? Do they have lights at night? Are the bugs very big? Do snakes crawl into your sleeping bag?' I assured him he would be safe — that with a little practice and listening he would find his way in the woods." She lightly touched the bark of an old-growth fir as we walked. "On his weekend vision quest, Renaldo really did meet a snake. But it wasn't poisonous and it was much too shy to sleep with anybody. The snake just coiled there — black with beautiful green diamonds. It blinked at Renaldo, stared at him for several long minutes, then graciously slithered aside to share the path."

"I said the magic passwords," Renaldo told Flor, his voice full of wonder and pride. "That snake understood."

Renaldo's totemic animal became the snake, symbol of power and rebirth. And because the snake's home was the woods, Renaldo was no longer afraid of the forest — he had an ally there. In school, Renaldo made a terrarium and bought himself a pet black snake. He said it was his science project. But he told Flor with a sly smile, "You and I know who this snake *really* is."

On every field trip, Renaldo's imagination grew more fused with the forest. When one of his ten-year-old friends was killed in a gang war, Renaldo began having nightmares of gang members breaking into his house and killing his family. Then in one dream, his snake appeared. "He grew big as a tree," Renaldo proudly told Flor. "And my snake scared away the whole gang." From then on, Renaldo saw his snake as his secret protector at home, too.

We stopped a moment on the trail. "Without nature we are all lonely children," Flor said softly. "Part of our loneliness and addictive behavior is that we have lost the connection to nature. . . . Maybe that's why even ten-year-olds turn to drugs. When a whole tribe of people is cut off from its source and crying for a vision, drugs are a substitute for what the Native Americans call 'medicine' or 'healing.'"

Above us the wind shook the tallest trees and white showers fell down on us like a blessing. As Flor and I walked back home through the snow, I thought of all the snakes still hibernating underground. I remembered my own early years in the woods. My first rattle was made from a rattlesnake's tail tied to a twig with leather thongs. As I crawled, I'd clutch that little rattle in my fist and so startle towering adults who might mistakenly step on me. When I was four and a half years old, we left the forest for some years by the sea. It was thirty-eight years before I saw my forest birthplace again. I returned to the high Sierra at the tip of Northern California to attend a week-long women's summer solstice camp.

At first it felt odd to be back in the same woods I'd known as a child. But after a few days, it was simple: I was playing again in my woods. Over the week two hundred of us attended open-air classes in meadows and among the trees. We sat in circles on the ground while we heard speakers on every subject from "Basque Mythology" to "Eleusinian Mysteries of Ancient Greece" to "Mayan-Hopi Wheels of Transcendence." We slept in tents or under the stars and soon the sensible camping shorts, hiking boots, and visors gave way to long, colorful skirts, bare feet, and bright gypsy scarves. It took surprisingly little time to forget the polite strictures of society and remember our more primitive tribal roots. In this, we had the eight-thousand-foot altitude as our ally.

On the last day of this gathering, I took my part in the Dance of the Tonals. It is said that when we are born, a *tonal*, or power animal, is born with us to live alongside and offer itself as messenger between Earth and spiritual worlds.

Led by a woman who'd spent her apprenticeship with a Peruvian shaman, our study group of ten women spent a day together in silence and meditation.

"You can call your tonals to come dance with you," the medicine woman said. "They will come gladly. They have never been far away from you. And once you remember them, they will always be here."

She told us that our tonals live alongside us like shadows, teaching us what animals know and humans have forgotten. "When you die," the medicine woman concluded, "the animal dies, too. And maybe next life, you trade places."

Then we did a series of exercises to call our animals home to us. Different tonals were silently summoned to take up their spiritual residence in each of the body's chakras, or power centers — from root to belly to crown of the head. In my navel center I felt the intricate circlings of a chambered nautilus; in my heart a grizzly; in my root and crown chakras two connecting serpents, coiled and patient as if they'd waited there forever for my memory to return to me.

"At first I didn't see anything," a woman said when we finally broke the silence to sit in one circle in the center of a stand of ponderosa pines. A thick blanket of dried needles cushioned us. Late afternoon heat shimmied in the air like a mirage. "Then all of a sudden I realized I wasn't looking up," the woman marveled. "I was gazing *down* at the tops of these trees. I was looking out the eyes of an eagle as it glided."

We all told of the animals who answered our silent calls. Next to me was a large woman who did have some of the gruff, maternal grizzly about her; next to her was a dove, several snow leopards, a dark-skinned jaguar, and a flaming-haired woman whose hooded black eyes gave her the regal fierceness of a red-tailed hawk. One woman I recognized from several of my other classes chose to keep her own counsel. She asked not to participate in the ritual painting and costuming that the rest of us undertook in preparation to dance in this ritual ceremony of recognizing and claiming our animals.

I knew this woman Diana's story; she'd confided it to me earlier in the week as we sat by the lake in full sunlight. Her story was set in the shadows: Diana was a double incest survivor. When her family's secret was discovered by a relative, she'd been sent to a foster home. There Diana lapsed into a silence that lasted several years. All that would rouse the child was sitting in her rural backyard forest alone for hours at a stretch. Even when Diana began speaking again, she never talked about her parents. She told people they had died when she was very young.

Now Diana had children of her own and a loyal husband. Still her hunched-over shoulders bespoke the hunted posture of the victim. When she spoke, which she did rarely, she cringed as if the sound of her own voice were too loud. Of course, the other women let her be. Of course, none asked Diana to dance.

In preparation for our dance, we busied ourselves in the forest. Grizzly looked for pinecones to make a great necklace, the Eagle feathered fallen branches across her arms for wings, the black Jaguar crouched low in a feline wariness. One woman emerged from her nearby tent with a leopard-skin mask she'd saved from her childhood. Another woman, Elephant, remembered to thud on the forest floor with her big, bare feet, pausing to listen to her own earthquakes.

At the sound of the drum's steady beat, calling us from the forest, we came back to make a circle in the pine needles. Many of us had painted our faces. I had silver zigzags running like lightning down my arms and face. Several women had elaborately painted bird faces, and one woman, Salmon, showed delicately etched blue-green scales up her bare back and legs. In the deep woods, without much clothing, with faces painted and bodies adorned by leaves and branches, anyone can become aboriginal. Anyone can remember that this is our native land and we are all primitives. The child in us remembers.

As I finished painting my legs, I was startled by a movement in the forest. Diana stepped lightly toward me, holding out a box of brightly colored watercolor crayons.

"Will you help paint me?" she asked.

Diana was so transformed, I just stared at her. Atop her head was a branch, a stretch of bark antlers. Her nose was painted black, proud and sensitive as she seemed to sniff the air for signs of hunters. Around each bare breast was painted a phosphorescent red-black-and-yellow bull's-eye. There was a flat, heavy stone clenched in each hand — her sharpened hooves.

"Who . . . what are you?" I asked, though I already knew.

Diana's voice was low. "A stag," she said simply. Then she turned around, naked except for a branch encircling each ankle. "Will you paint my ass like my breasts?" she asked.

I didn't move. A fear came over me. "I can't. . . ." I said. "It's . . . it's too terrible." I shook my head. "I don't want to help make you a target again."

Diana fixed me with dark, oval eyes that softened slightly. "Please," she said. "It's part of my dance." Then she laughed huskily. "I can't be my stag without it."

I nodded. As she stood at her full height, I painted two bull's-eyes on each buttock, my hands shaking.

In the circle, the drum did all the talking, like a great heartbeat in the forest. After dancing together for what seemed hours, we all sat silent. Then each woman stepped alone into the center of the woods and, in rhythm to the drum, let her animal move with her.

Grizzly never left the earth; Eagle never touched

down; Snow Leopard danced only a second, then as usual eluded us in the underbrush; Elephant broke the silence with a scream and shook the ground with her stomping two-step. I did my Cobra dance low in the pine needles, swaying to the drum as the trees above me swayed. On my belly, I felt the Earth pounding against my navel.

At last it was the Stag who stepped into our center. A few women caught their breaths, hardly recognizing Diana. She had never moved like this: deliberate, forceful steps, massive head turning this way and that to watch us, eyes black like bullet holes in her impassive face. As Diana danced, a masculine sway to her wide hips, her hooves pawing the ground, I remembered a hunting scene I'd witnessed once in Colorado.

In a game preserve, I stood watching a herd of deer graze. Suddenly a truck pulled up and some drunken men jumped out. They were not hunters; though this was the season, they wore no orange jackets. They had no notion of tracking, the stealthy forest stalk. One man rested his rifle on the barbed-wire fence of the game preserve. Before he could get his aim, a subtle change ran through the herd. As if electrically charged, they all grouped together tightly and moved slowly backward. From their center stepped a huge buck; his antlers must have been six feet across. Very deliberately, the buck stepped forward, steadily moving toward the man, his head lowered. With a curse at what he thought was the buck's challenge, the man pulled the trigger.

The shot echoed off the far ridge. Like gazelles, the herd scattered swiftly. But the buck kept coming toward the man. Another shot. The buck dropped to his knees, stared straight at the man, and toppled. With a hoot of triumph, the men jumped back into their truck and took off. There was no ceremony, no asking forgiveness of the deer for his sacrifice, no ritual to clean and dress the buck, then partake in his great spirit so he might live on in our nourished bodies.

I crawled through the wire fence and ran toward that buck. It lay alone in the meadow. But I could feel the eyes of his herd watching me from the forest. I kept my proper distance; the buck was still alive. He lay there bleeding from two wounds, panting, his eyes liquid and dilated. At last the dark eyes fixed, rolled back. With a breath like a sigh, the deer died. As my father had taught me, I put a branch in the buck's mouth — food for his journey to the spirit world's forest. Then I laid my palm on his warm flank, tracing with one finger the bloodied bull's-eye.

That same stag danced again in Diana. She was all thunder and rage as she spun around within our circle. Some women fell back from this raw display; other women leaned forward, eyes riveting on the fierce antlers adorned with dangling brass earrings. The Cobra in me swayed as the ground echoed staccato poundings of drum and fading hoofbeats.

Much later Diana rejoined our circle. She still kept her own counsel. But I never saw her cringe again, not once.

The next day when we were all leaving that forest to return to our homes, Diana approached me quietly. "Thank you," she said, her eyes steadily holding mine.

"Yours was the most beautiful dance of all," I told her.

Diana threw her head back, then said in a deep voice, "The stag in me was never wounded." She laughed. "They missed me. They'll always miss me." Diana turned to leave, calling out, "Goodbye, Cobra."

I have never since then seen a deer without thinking of Diana dancing naked in the woods, bull's-eyes all over her body. On my walks in the forest, if I see a snake slithering across my path, I think not only of my own totemic animal, but also of little Renaldo somewhere there in the wilds of Los Angeles fending off gangs like the rattler warns his predators.

Now when I find myself midlife walking in the dark woods, I know I am not alone. The animals are my allies; the trees are gods and goddesses who in deep stillness keep the Earth's counsel. All that is alive calls out to me to come play, to take my part in the dance.

Moose Man

*T*he first time I met Joseph Meeker, I was alone, hiding out in my office, AWOL from the daily wars. For half a decade, I'd picked my picaresque way between the pitched camps and philosophies of the environmental movement. Movement is hardly the word for those of us mud-splattered and bedraggled in the trenches. It was more like the War of the Roses, the Trees, the Spotted Owls, the Whales, the Old Growth, and the Wild Rivers. In Meeker's own words, the debate over "whether nature belongs to us or we to it."

I was weary of this war and confess to mutinous thoughts, such as suspecting that the hard-line environmentalist camps sometimes seemed as humorless, self-righteous, and willful as their governmental adversaries, the Bad Guys of the U.S. Forest Service, the Bureau of Land Management, and, of course, everybody's favorite — the man who intended for the land what Hitler did for the Jews, James Watt. But that morning in 1986, I was so weary, I found myself remembering James Watt somewhat fondly — his exit from the Department of the Interior signaled one of the few hiatuses in the environmental war when both sides at last shook hands in perfect agreement over this common enemy. Then back to business as usual: the snipes, the strategies, the lobbies, the laws. As usual, I was in between, surrounded on all sides.

My job that day was more impossible than usual. I was to write a brochure acceptable to my employer, REI, several environmental groups, and the U.S. Forest Service. It was on using, maintaining, and preserving national forest trails; yet it was somehow to avoid the deep controversy between environmentalists and forest service over those same trails, or lack of them. I sat there over my third cup of herbal tea, telling myself that if I could give up coffee, I could certainly give up this daily adrenaline rush of trying to write my way between the firing lines. Like so many civilians trying to get by in a war zone, ducking out (or desertion, as the military insists on calling it) was one of my favorite survival strategies.

So I put my feet on my desk, declared it teatime, fixed a cup of Sleepy-Time tea, and indulged in my deepest treason: instead of reading *Wilderness* magazine cover to cover and taking notes for my job, I just looked at the pretty pictures, in the way as a child I'd wander through *National Geographic* before I could read.

So at teatime, lost in a stunning waterfall, my eyes wandered and fell on familiar words, much the way the eye absorbs graffiti unconsciously. And there it was again, the war. As I read, the references to "The Chief" this and that sprang like a booby trap inside me. "The Chief dreams of a day when we all love the land and, perhaps even more importantly, will cease making the life of the Forest Service so difficult." Then I remembered: I was not exploring a world without wars, I was just resting and tea was my truce. I was not an Indian girl sunning herself by some forest waterfall. I was in an office. Just like my father. And he was not an Indian chief looking out over his land before the white man made war on it; he was the chief of the Forest Service, whom this article took to task.

For the eight years my father served as chief of the U.S. Forest Service, I worked as an environmental writer. When he retired, I used to joke, so could I. And that did happen eventually. But that spring day back in 1986 when the environmental wars were at their peak, when I wanted a rest more than I wanted to be right, I gave up. In fact, I did what many people do, because war can be quite boring. I fell asleep.

Everyone was going home and I was known for my odd hours, so no one knocked on my door. When I did rouse myself, it was late evening and the office scene had changed to the scenario I liked best — me and the janitors. No phones, no chat, just the lulling roar of the vacuum and the canary-whistle of a lone security guard. Now this was a time when I could really get some work done. I gathered up the *Wilderness* magazine again, turned to that page about my father the chief. I read the conclusion: "At stake, as it always is, ever and ever, is the land and all that it holds of what we have been, are, and will be. So the buttons are pushed and the plans come bubbling up out of the computers, and the environmentalists take them on, take them apart, appeal them, sue over them, with or without confidence what they do can force a change in the monolith. . . ."

I stifled a yawn, put on the kettle, and rolled a small piece of tree into my typewriter. This time I'd do it. This time I'd write something that would speak to all sides, restore some common ground — after all, one thing everyone concerned with the environment agrees upon is love of the land.

Then something wonderful happened. I made a mistake. And from that mistake my world shifted. I was leaning over the magazine, looking for a quote, but I'd turned to the wrong page. On the opposing page — how appropriate — was an article entitled "Some Earthly Speculations," by Joseph Meeker. I was struck first by the pretty picture, a drawing of a man whose head was full of trees. Then I read: "Like the deer, I need not travel in straight lines with my trail. It is a principle of forest life that straight lines do not exist." I started to laugh. This was my kind of trail! And the deer delicately picking her way through the forest fires of an environmental war seemed to offer a natural survival strategy for the animal in me as well as the writer.

I read on: "As I walk the trail," Meeker was describing his handmade trail on his own ten acres, "the forest is making tracks upon me that will not be easily erased." Again, I smiled. It was such a simple shift — from humans managing or preserving nature to nature having her own way with us.

Meeker concluded, "The universe has been quite literally writing upon humans for many thousands of years, and our alphabets are among the trails that nature has carved in order to cross our minds. Wild lands have cut deeper trails in my life than I will ever be able to make in the forest."

"Aha!" I said out loud. Then I perched on my knees at the typewriter and wrote that little trails brochure in a dash. Well, it worked. No side took offense, none defense. And probably of all the writing I'll ever do, that little trails brochure and another on wilderness water might well be the most widely read.

That night when I finally turned off my typewriter and headed home, I vowed to do what I've only done once before: write a fan letter. I did. It was received by *Wilderness* and passed along to Meeker who, it turned out, lived within spitting distance of me, across Puget Sound on Vashon Island. Then I met Joseph for the second time.

This time it was not his writing, it was his voice. Mellifluous and deep and musical like a walking bass, he telephoned and asked that I call him Joe. Then we were off on our first rollicking mutual recognition that here were kindred spirits. His background is comprehensive: he's worked as a national park ranger in Alaska doing wildlife studies, taught comparative literature and philosophy in academia for over sixteen years, and he's shuttled between the separationist camps of government and corporations. He's also had his own ecology program, "Minding the Earth," on National Public Radio. Here was a veteran survivor — he didn't take sides, he learned each side's language and then translated the best of them. Here was someone who for fifty-odd years made his light-footed way in the no-man's-land between the longtime forces of scientific scrutiny and storytelling.

Whether it was teaching a whole-systems course at Antioch on "Natural Boundaries," or lecturing at Scotland's Findhorn Institute on "Nature and Other Mothers," or writing essays in what he calls "the comic spirit . . . a serious form of play with the earth," Joseph Meeker is a practiced fringe-dweller. I could learn much about survival from this man, I thought. And we two, both born in the sun-drenched month of August, settled on a date for our first of many "Leo lunches."

I remember praying before that first lunch that Joe would be ugly, as narcissistic as some visionaries who are so self-absorbed they naturally put one off. This would protect me from falling in love with him as I had his work. But Joe was none of these. Gracious, open, playful as an old Taoist monk, Joe looks like a very tall troll or a moose. Though his eyes are blue, they seem hooded, almost Chinese, and his forehead stands out like a bald promontory, across which run wrinkles from years of weather and cigarettes. He reeked of tobacco. It is impossible to adulate Joe; he's so pointedly and imperfectly human. No guru, rather a trickster or as he describes in his book *The Comedy of Survival* (Charles Scribner's Sons, 1974), the picaresque character, quite the contrary of the tragic hero. The tragic hero must die to fulfill some moral order that is supposedly higher than the immoral natural world. But the picaresque character's strong knack of survival in life, if not in history, is always the same: "Adapt to circumstances and take evasive action."

In his inventive and original *Comedy*, Joe marries Hamlet to the animal kingdom, comedy to our human biology, and reconnects sky to earth. The book begins:

> Tragic periods try to do good and save the world; comic periods try to do well and to encourage the conditions supportive of life. In tragedy we learn to hate the evils of life; in comedy we seek joy from its goods, scarce though they may be. . . . The period that we are slipping into now looks more likely to be one of disaster than of tragedy. . . . As awareness of disaster grows, we can expect to find ourselves worrying less about moral purity and more about our responses to immediate threats. Perhaps we will spend less time trying to transform the world and more trying to change ourselves to fit the world.

When I first read his book in 1986, it seemed eerily prescient. Now from this vantage point of a new decade, with earthquakes in unheard-of places, all-too-active volcanoes, Berlin Walls falling, and massive upheaval worldwide, the book has never been more timely.

First published in 1972, *The Comedy of Survival* asks us to consider learning to live without the terrible, destructive consequences of the tragic hero myth — and like true comic survivors seek or invent alternatives. After all, Meeker writes, "however the human mind imagines the world, that is how the world tends to become." If our stories, our sciences, and our saviors speak of an evil world populated by sinful humans who must die to transcend its natural limitations, it becomes our fate to destroy our Earth.

But the comic hero, as Meeker puts it, "is so completely absorbed as a participant in life that it never occurs to him to be a critic of it, or to escape into fantasies." The comic hero instead is just a guy or gal trying to get by and get along with the world, as well as with our fellow creatures. The comic in us may bumble and fall, may succeed without much elegance or eloquence — but survival, not death is what makes us truly heroic. The comic hero pays close attention and adapts to rapid change, is flexible, finds happiness in the smallest details of daily life, avoids pain, and claims kinship with everything from amoebas to elephants. Disdain, contempt, self-righteousness, and conquest of man or nature belong to the tragic hero's drama. The comic hero is much more likely to find any way he or she can to shake that enemy's hand and ease on down the road.

In this last decade of the century — in which the tragic hero in us has discovered how to martyr us all by destroying not only our enemy, but our species — it's heartening to have a voice like Joseph Meeker's crying out in the wilderness. "Given the opportunity," he writes, "people have always shied away from killing their own kind, in spite of all training to the contrary." Meeker suggests we might do well to imitate the animals, who have always been more interested in survival than killing their own kind. Even predators take from their prey only what they need. Lions don't stockpile dead antelopes, and eagles don't ritually murder an entire run of salmon because those salmon violated territorial treaties. As Meeker concludes in *The Comedy*, "Animals now seem more likely to enlighten us than angels, and nature more than God."

Another form of human enlightenment that helps us adapt to rapid change, laugh at ourselves, share simple pleasures of ordinary life, and keep us in balance is what I'd call the mutual comic heroism of lasting friendship. "Companionable participation, not power, is the best gift we can offer to life," Joe wrote in his latest book, *Minding the Earth* (Latham, 1988). In the five years since Joe and I first fell into friendship, we have companionably participated in one another's lives and work. And like any good friends, we have come to recognize one another's contradictions. As Joe himself said, "Belief is one thing; behavior another."

Joe is the first to remind me, mid-drama, that I'm taking myself too seriously. One Leo lunch as I furiously described an editor's high-handedness with my work, Joe sat back, grinned, and suggested, "Why don't you nuke him?" When I stopped flailing my arms and stared at him, openmouthed, he shrugged, "Just a thought."

Another time when Joe and his mate and I and mine all hiked the Monterey cliffs, I introduced Joe as "a true environmentalist — one who'd rather pollute his body than the Earth." To which Joe laughed and lit up another cigarette. Then we all sang "Darktown Strutters' Ball," as we drove the stunning curves of California's Highway 1.

But a few years ago, Joe and I didn't laugh much over my favorite fond nag of him. And a few years ago, Joe had the misfortune to endure what many writers try to escape: living out their own theories.

"How dare you be a tragic hero after persuading me to give it up?" I demanded, in spite of myself, when I heard Joe's familiar, deep, lullaby-voice on the phone. But he wasn't singing; he was telling me that the next morning he'd be in surgery. Throat cancer, just discovered. They'd have to take his jawbone and part of his tongue along with the malignant tumor. He might lose his voice, have to learn to talk again.

"No, not that beautiful voice," I protested. I felt it was as much my loss as his. I didn't want to go back to a time when I only knew Joe through the written word, when we couldn't laugh over lunch and rib one another. A voice like Joe's could read the phone book and I'd listen. "This is *not* the way the story is supposed to go, Joe." I told him.

"Well," he said softly. "Maybe I need to be quiet now. Maybe I've said enough." Then he paused and laughed — a throaty, full guffaw. "Maybe it's your time to talk. You haven't given a reading in five years. Go on . . . give one for me."

There was absolute silence as the full realization hit me: Joe was saying goodbye. I might never hear him again. There was so much love filling that silence — it was like another voice.

"Will you think of me tomorrow?" he said.

"I'll do nothing else," I said. "But Joe, can't we find some way to turn this into a comedy?"

"Of course," he said. "We'll all just have to put our heads together."

That next morning, as Joe tells the story now, there was a wildly ecumenical congress gathered in spirit around Joe's operating table. "There were messages from Celtic Christians and from skilled secular mediators. Druids and Taoists sent their spirits, as did witches, Zen Buddhists, and Hindus," he reported. Also offered were the more orthodox Catholic, Jewish, and Protestant prayers.

I sat in my office, door shut, in the very spot I'd first read Joe's article, and closed my eyes. What I remembered was Joe in his own backyard forest, head thrown back, laughing as we contemplated the new outhouse he was building from his fallen timber. He named it the Friedensreich Hundertwasser Memorial Meditation Center after the artist/architect who believed that if we blessed our food, we might also remember our excrement was equally sacred. The memory changed, shifted, became dreamlike. Instead of in his own woods, Joe stood under a waterfall like the startling azure gush of blue water from rock in Arizona's Havasu Falls, a place Joe and I had both visited, but not together. I imagined this water cascading down through Joe's open mouth. He was laughing, he might as well take in water. Somehow I hoped it would purify.

All over the world, Joe's friends must have had similar images, and Joe himself remembers the surreal surgical sequence quite vividly. He felt his friends very near him. When he awoke, he was told that the tumor had been mysteriously and completely encapsulated. Surgeons took it all, but no bone, no tongue. His voice was not stilled.

Joe still smokes, of course. Still reeks; still smiles when I tell him I take great satisfaction in the fact that fate did not allow him to take the tragic hero's way out. He has to keep living like the rest of us schmoos in this comedy of survival. Someday, he threatens me, we'll find ourselves doomed to writing a book together and that will make short shrift of our friendship. Then we can rest and be lifelong enemies. We can carry on some comfortable feud even when we're senile and in the same senior citizens center. We'll call it the Black Hole Nursing Home for Wayward Writers. Until then, there are our Leo lunches. Recently we invited another Leo as a guest to one of our lunches — my father. I told him before we took the ferry to Joe's house, "Please, Dad, don't try to tell Joe how to manage his forest."

My father laughed and pulled a deep draught on his pipe. He's retired as chief now and I've left my environmental-wars writing job. Now my father is an advocate for international fish and wildlife agencies; he has emerged as a thoughtful environmentalist. There is less and less we disagree on — though our approach to nature will always be different. He sees himself as a steward of the Earth's resources; I suspect the Earth is so alive she'll survive with or without us. "Don't worry," my dad assured me before our lunch with Joe. "I never tell a man what to do with his own land."

"Joe doesn't think the land is his," I countered.

Dad sighed and puffed his pipe. I decided with the two of them smoking, we should sit outside. "Whatever . . ." Dad said and let the smoke waft eloquently between us.

I recognized his tactic. He was evasive, like any true comic hero. And I smiled. I remembered once my father told a congressional committee that testifying on controversial forest issues was not as difficult as raising daughters.

"Why do you call him *Moose Man*?" my father asked. He knew a story always makes me forget issues.

We sat back and let the ferry rock us, water splashing high against the white sides.

Well," I began, "when Joe Meeker was in Alaska studying moose with the Park Service, he was following this beautiful moose."

"As big as the old guy I got?" my father asked.

I remembered those moose antlers spreading across our family fireplace. I used to watch them more than the television, there was so much life still in them. "Probably bigger," I suggested.

"I doubt it," my father pursed his lips and grinned, clicking his pipe between his teeth.

Then I told him the real story, as Joe had told me. This moose was a mother, it turned out. And Joe made the big mistake of coming between her and a new calf. In a second, the moose lowered her head, charging Joe full force. He scrambled up a tree and stared down at her as she snorted and pawed the ground. Joe said her eyes were bloodshot with rage and white clouds puffed out of her nostrils. She was one mad moose and there he was, a rather hapless character six hours up a tree in the dead of an Alaskan winter, with no buddy around.

"Frostbite?" my father asked.

"Almost," I said. "Then a strange thing happened. Joe was looking down at that moose and she was staring up at him for so many hours that they got kind of used to gazing at one another. Joe said when he'd look away, he'd get kind of lonely. Then he felt kind of dizzy and realized . . . well, he'd fallen in love."

My father shot me a glance, but said nothing.

"That's what it felt like, anyway. Love. The moose must have felt it, too. Because she just stepped back, still gazing up at him. He knew it was safe to climb down the tree. And when he did, she just stood there, still staring at him. He thanked her and she let him pass. Between them was complete acceptance."

"Or maybe they were just tired of looking at one another," my father suggested. But he smiled.

When my father met Joe Meeker, he found a moment to take me aside and comment, "He does look a little like a moose, now . . . doesn't he?"

We had lunch at Joe's home on Vashon Island. Dad and Joe talked about Alaska, but discreetly avoided any environmental controversy. They seemed to be sizing one another up, much like that moose and Joe did so long ago. Joe suggested we all saunter on out into his backyard ten acres and walk that trail he'd hewn — the trail about which I'd first read.

We walked three abreast along the beginning of the trail, Joe pointing out a chanterelle mushroom here and there, my father commenting on the variety of trees, from fir to madrona to some spindly cedar barely making its way up toward the light. Then the trail narrowed and we all walked single file, wrapped in a silence like an accompanying mist.

"You know, Joe . . ." my father began, lighting up his pipe. He raised his eyes. "You could thin a few of those trees and thicken up this forest."

Joe stopped, considered a moment. He lit a cigarette and cleared his throat. "Well, Max . . ." he said, his voice never more melodious. "I don't manage this forest. I just let it be."

My father opened his mouth and I prepared myself for another debate, perhaps another environmental war right here in this quiet backyard forest. Then my father grinned and nodded. "That's right," he said. "That's another way of doing things."

When my father returned to Virginia, he did something I'd never known him to: he wrote a bread-and-butter letter to thank Joe for his hospitality. In my family, those letters are usually left to the women.

I like to think that if a moose can teach a man new tricks, then maybe we can teach ourselves new ways of "companionable participation," with people and other animals.

There is a Skagit Indian myth that tells of a common language spoken by humans and animals and nature. It prophesies: "When we can understand animals, we will know that the change is halfway. When we can talk to the forest, we will know that the change is come."

Like any comedy that ends with characters in a marriage of true minds who speak the same language, I like to think of this meeting between a moose man and my father as a small sign of this welcome change to come. And who knows what part the forest, the sea, and the animals have yet to play in this earthly comedy as we reconcile ourselves with this world, the only home we have ever known.

On Drowning

*A*t the age of two, I saw the ocean for the first time. I threw wide my short arms and ran shouting, straight into the Pacific, where an undertow reached out to embrace me. I still remember the spinning upside-down whirlpool of warmth like the womb I'd so recently swum my amphibious way out of.

Spun round in the waves of that undertow, I remembered my first mother. All my air gone, I instinctively opened my mouth like a guppy, breathing in the dark, nutrient-rich liquid. There was no panic, no struggle. I tucked myself in tiny somersaults and suckled seawater. Who knew which way the sun shone? The ocean had its own light: through bulging eyes I saw bright blowfish and bioluminescent bacteria shine through the water like a sea-bottom constellation. There were pink coral and purple sea anemones, sea cucumbers slithered past as my fingers sought their velvety, speckled backs. Blazing orange starfish inched past snaggletoothed, slinking eels — and just when I closed my eyes to rest from such undersea splendor, I heard the far-off lowing lullaby of a gray whale singing me to sleep.

My father woke me, rather rudely. Cranking my arms like an old-fashioned Model-T and thumping my chest, then sucking the sweet seaweed from my mouth, he kept yelling at me — the way he did when it was time to come in from my play. There were lots of people around. I was afraid to open my eyes, which stung with sand and tears. But I heard the crowd, skittering around like sandpipers and calling my name as if I didn't know who I was. What I remember most is my anger at being brought back home before I wanted to return. And the dawning, disloyal conviction that my real home might be this ocean.

"You could have drowned!" My mother was beside herself.

"You did drown," my father said grimly. Then he crushed me to him — an experience more physically painful than the undertow.

But I allowed it because my parents had forgotten to punish me for running away from them. "Pretty," I assured them. Then I tried to explain about the fishes that made their own light and the underwater world that opened up to me.

"No, honey," my father gently dismissed my vision as if I were in a delirium. "You didn't see anything, just a lot of silt. You weren't deep enough for those bright little fish." Then he promised me, "And you'll never go that far down . . . at least not until you can swim!"

But I did go down, I still do go down to the very bottom of the ocean — though not physically. Ever since that so-called childhood drowning, I have had dreams just like what I witnessed in that long-ago undertow. The dreams have the opposite effect of despair. Instead, this dreamtime landscape exists alongside my daily life as companionably as water lapping its own shore.

Many nights I find myself sleeping on the ocean bottom, my body drifting in the fluid pulse of the sea. In dreams I recognize almost casually that I'm not human, but aquatic with all manner of gills, fins, antennae, exoskeletons. In one of my favorite dreams I was an octopus with red suction cups, jetting through the water and squirting ink in my elegantly elusive wake. A dream I usually only tell children is of living inside the belly of a blue whale: I'm just so much silly krill slipped through the blue's mouth. Children, knowing the power of being small, don't mind imagining themselves as such tiny shrimplike crustaceans. And the innards of a whale seem no odder a residence than the recent round womb of one's own mother.

Early on, I learned to keep mum about my drowning dreams. When I was five and my great-grandmother died, I tried to ease my mother's sorrow by suggesting that Great-Grandmother was back home in the sea — which earned me solitary confinement sans supper. Such sacrilege in a Southern Baptist home was not taken lightly. Had Satan put it into my head that the Earth's own ocean was heaven, not — as everyone who was saved knew — the sky?

At eight, when I should have known better, I asked a Sunday School teacher why heaven was up and not down. I was referring to the bottom of the ocean; she suspected I spoke of hell and took me on.

I mistook her missionary zeal for openness and blurted out, "Maybe heaven is at the bottom of the ocean. In my Bible's dictionary there are thirty verses under *sea* and only two under *sky*."

"Yes," said Mrs. Lucas, with a look of evangelical triumph. "But now let's look in our Bible concordance under *heaven*." Dutifully we all did. It filled up half a column. "And heaven is far above us. It's someplace to which we *ascend*." She rested her case and busied herself explaining ascension to Billy Marshall, who always flunked our local school's second-grade vocabulary tests.

I was not convinced. When I read Genesis 1, all I saw was that God said, "Let there be a firmament in the midst of the waters." When I looked up *firmament* in my school dictionary, it explained that this was the "vault of heaven," and though the first definition listed for *vault* was an arched ceiling or roof, the second was "an arched space, chamber, or passage, especially one located underground." A final definition was "burial chamber." I was thrilled to think God's firmament in the waters was a secret passage or burial chamber at the bottom of the ocean. Maybe when I died I'd take up my heavenly residence inside one of the many pearly mansions of a chambered nautilus?

I kept this idea firmly to myself and did what many pagans do with Christianity — I adapted it to my private mythology and secretly still worshiped the old ways. I kept these dreams and notions even from my siblings, which was no mean feat because we all shared our secrets.

My undersea secret life was continually nurtured by the fact that my father almost always located us next to or near water. Walking or playing by water was daily worship for me — after all, what went on by the water was holy compared with the pounding, snoring, fidgeting, and downright droning atmosphere of church. In third grade when we lived across the street from the Atlantic Ocean in Revere

Beach, Massachusetts, I discovered a socially acceptable way to speak of my underwater world: Earth Science. It was my salvation to give my siblings long discourses on amphibians, those near-kin who begin as water beings and then grow into land dwellers.

"We were all amphibians once upon a time." I told my younger sisters and brother my scientific discoveries like bedtime stories.

Gleefully I'd show them my Earth Science book with its colorful illustrations of green frogs and coral pink sala- manders. I'd remind them of the boy in our church born with webbed feet. "This is what you looked like inside Mom," I'd tell them. They'd gape like little fish. "You had tails and gills and breathed water just like a tadpole."

Anchoring my science book in my lap, I'd pretend to delve into deeper chapters of scientific fact. But what I'd really tell them were my dreams and stories. I'd read of the marvelous mer-people, a tribe of undersea people who lived in vast caverns — "heavenly vaults," I called them. I showed my siblings *National Geographic* photos of Anasazi cliff dwellings as dry-land evidence of how ancestors of the mer- people lived their sea life in caves before the waters receded.

Each dusk when my sisters and brother and I combed the cool Atlantic's summer beach, looking for couples entwined like so much tangled seaweed or drowned people cast back on shore, I told them my favorite stories of sea changelings: the undines, selchies, sirens, and water nymphs who wander between the worlds. I didn't tell any- one I secretly suspected I was one of these changelings.

Aside from the mermaids and mermen who swam through my dreams, my favorite changeling was the selchie. In the sea, she is a sleek, playful seal; on land, she's just like you or me, except maybe more fun. We'd seen sea lions and seals on our trips out West and the vision of those flapping flippers, the sheer speed of their torpedo bodies slicing through waves, reminded me of the weightless acrobatics I performed nightly in my dreams.

Often my siblings clamored to hear the South American tale of the ancient pink dolphins who, the Amazon Indians believe, navigate their nightly way up shallow tributaries. Leaping ashore, the dolphins change into lovers, reunite with their human mates, lie with them wrapped in the rocking waves of dreams. From this sacred bond are born children the tribe considers half-human, half-dolphin. These children are special go-betweens or shape-changers who can heal the broken treaties between tribes of land and those of water.

The only story I didn't tell my siblings was of the undine — which for the longest time I thought was spelled *undying*. The unfortunate undine was a water spirit who had to marry a mortal man and give birth to his child. Only then could she be granted her soul. This injustice and Hans Christian Andersen's wrenching tale of "The Little Mermaid" made me so mad I boycotted them. My youngest sister also read Andersen's moralistic little tragedy of the little mermaid who falls in love with a human prince, saves him from drowning, only to witness him mistake his savior as the first young woman who finds him on the shore where the little mermaid has so lovingly laid him. The mermaid sacrifices her beautiful voice, her undersea kingdom, her sisters and father — all to dance on human legs that cut her like knives. She devotes herself to the prince who dotes on her but marries the woman he believes saved his life.

What enraged me most about the story was Andersen's insistence that mer-people had no immortal soul. In his version, undersea folk lived for three hundred happy years then changed into so much sea foam. Human beings, however, ascended — there was that word again — to heaven to live forever with God. After all her longing and sacrifice, do you think the Little Mermaid got her soul granted by marriage to mortal? No. She's given a chance to redeem her three hundred years of undersea life by killing the prince on his wedding night. She flings the murderous knife into the sea, then herself changes into foam. But wait! She is not just nothing. She's granted a place among the "daughters of the air," who through good deeds (mostly making bad children behave) earn immortality and "share in mankind's eternal happiness."

"So who wants to be a mermaid?" my mutinous sister said, after reading Andersen's story. "You'll only get to heaven if a mortal loves you or if you go into the air and do three hundred years of good deeds. You'd never last that long trying to just do good."

This was my difficult sister, the one who often doubted me. Now she was old enough to flaunt her incredulity. I was twelve and she was eight. "You're just like that Little Mermaid," she said scornfully. "Once you're here, you want to go back home. But you never can because you didn't save your soul while you were on land."

"'The Little Mermaid' is a made-up story," I countered.

"You make up all your stories, too!" my sister shouted and flounced away as if I were so much dead sea foam at her feet. "Liars don't go to heaven."

"Well, then, I'll stay here!" I called after her. "I won't be lonely."

"Sure," my sister said matter-of-factly. "You won't be lonely here on Earth, you'll just be lost."

There were worse things than being lost. Being an adolescent, I knew this well. It was also during this time I began to understand there were other kinds of longing than my dreams of the sea. There was the same longing that Little Mermaid endured — for the beloved human.

I never came close to sacrificing any body part for my first boyfriends, but I remember a girl in ninth grade responding to one of my Creative Writing exercises about my mer-people by saying sadly, "Whenever I think of mermaids, all I think of is suffering . . . the terrible price a woman pays for love."

This girlfriend played in the symphony with me. She had a crush on the first clarinetist, and I was really gone over Johnny Rodriquez, he of the soulful alto sax. He looked like a junior-high Ricky Ricardo, and when he stood up to play astonishing riffs on his sax, even the symphony director stopped scowling.

Sitting forlornly three sections back as a third clarinet, I'd listen to Johnny's solos and feel as if I'd lost my voice. That whole year I never spoke to him once, even though twice he walked me home. I heard via the symphony grapevine that he called me "catatonic with clarinet" because the cat had so got my tongue.

I was beginning to believe my sister and Hans Christian Andersen were right. This fear and the fact that we'd moved to Virginia right next to our church's preacher and spent almost every night now in that fundamentalist church sometimes even set me to wondering about my eternal soul surviving an undersea heaven on Earth. Then, just when I was about to lapse and entertain ideas of ascension or joining the adolescent daughters of the air, I drowned again.

Because I'd been breathing air for twelve years longer than the first time the water took me, the transition was not so easy. Our family was moving again, something we did every other year or so, and we were mid-country in a Nebraska motel. After a day of steaming inside a station wagon, our family of six splashed into the motel pool, even though it was night, with no lifeguard and no pool lights. For all my many nights of undersea explorations, I am not a strong swimmer by day. I've never learned the ubiquitous crawl, my basic stroke being a simultaneous frog-kick with legs and lunge with arms to scissor my way through the chlorinated waves.

Since in my dreams I breathe water, holding my breath requires an effort of will, something I don't associate with being in the element of water. So I'm always somewhat sadly disappointed when I swim in pools. That Nebraska night was no exception. Of course, what no one mentioned during our water play was the grief we felt at leaving yet another homeland.

My father played a dunking game with us: just as we surfaced from our deep underwater dives, he'd let us gasp once, then push us back down below. With no pool lights, my father didn't see that I surfaced, forgot to breathe or else got confused and breathed water. With a shout, he put both hands on top of my head and in a mighty shove pushed me down to the bottom of the cement pool. He held me there, counting aloud while everyone laughed. It was too dark to see my last air bubbles, too noisy to hear my screams.

I remember sitting cross-legged on the bottom of the cool cement pool, my father's hands cupping my head. It was unfamiliar, this panic, this tearing pain in my chest. I screamed again and then inhaled a great gasp of water. There was a split second of complete lucidity. *Oh*, I thought, and was suddenly grateful, *this again.* Then the calm as my body returned to its watery rhythms. But this time there was no splendid sea bottom. There was just darkness and a distinct loss of feeling in my legs.

I woke with a howl as someone popped open my sternum bone like breaking a lobster shell. My father's face swam before me. I had two bodies — my fragile, bony chest and arms while below the waist my legs felt fused — somehow a familiar feeling. In one powerful leap I thwacked my tail and saw it shining with translucent blue-green scales as brilliant as mica. Then I passed out again.

When I came to, I was stretched out on a pool chair,

towels wrapped tightly around my body. Everyone around me was dressed in white towels, too, like bathers in an ancient Greek underground gymnasium. A pool gleamed in the moonlight. I was in another world until someone handed me a Coca-Cola and I heard my father's voice.

"Not again . . ." he said wearily, his face very white. "I almost couldn't get you back."

"Yeah," my little sister piped up. "Next time you try to drown yourself, Dad might not be around."

"She didn't do it!" my other sister and brother angrily defended me. "Dad did . . . except he was only playing."

"Did you see heaven?" my little sister demanded. "Well, did you?"

"No," I said slowly, my ears still plugged. "It was too dark to see. Except I did feel my legs change into . . ." I stopped.

My little sister gave me a meaningful you-know-what-that-means look. She was merciful enough not to mention hell or how the Devil had a tail, too. However, she did remind me that when God first destroyed the world he sent water.

These were the dark years of disbelief and forgetting who I really was — in other words, that long plague called adolescence, that time when one's first world is destroyed and who knows what comes next. "After the Deluge," I call these teenage years, because it was then I began reading the myths of lost continents sunk fathoms below because of the so-called sins of their advanced civilizations. I didn't like the sin stuff, a reprise of all I'd learned in church. But I was fascinated by this deluge myth common in so many far-flung cultures. My favorite was the Toltec myth adapted by the Mayans and Aztecs to tell of the five worlds the People journeyed through to come to this last world. In each world, the gods of air, fire, water, and earth look down and see the People have forgotten how to give thanks for their world.

Loath to completely destroy the People they created and love, the elemental gods search for one good man and one good woman. Of course, those who still believe or remember to sing praises are rare, but at last the one good man and woman are found. They are given a cypress tree for a raft and to this, as well as their faith, they cling while the thunderclouds cover the world, the volcanoes spit long tongues of lava, the earth trembles from its fiery core. All the world is drowned, all the wide world is water, except for one volcanic peak.

As they are drowning, sinking to the bottom of the great sea, the People remember their gods and call out their names with their last breaths. Mercifully, the gods change the lost People into all manner of brightly colored sea life — fish and great whales and all the seafolk that swim and breathe life in the deep. All that is left of the once vast tribes of the People is one good man and woman afloat on their cypress bark. When the waters recede, this one man and woman begin again. And as the story goes, they have plenty of fish to eat.

After my own adolescent Deluge days, I also re-created my world: I left home for college. I left behind the first world of my parents' beliefs and their suspicion that I was absent among the saints: those saved and destined for the sky. At school I studied comparative literature and dove into world mythology with my old passion.

Ovid's *Metamorphosis*, with its myths of old couples who turn into trees and lovers who fling themselves off cliffs only to change midair into sea gulls, was another bible. There were the ancient myths of the sea goddess Thetis and later the Greek sea god Poseidon, who also rules the underworld. In these myths were water nymphs galore and a wise Venus rising from the waves, one delicate foot on a cockleshell. I rediscovered that even in early Christian symbolism those first followers of Christ used the sign of the fish, fish being a ritual religious meal; they celebrated their new belief by baptism — immersion in waters holy enough to cleanse every sin.

When at last I arrived here in the Northwest, I sank myself into a study of native Indian tribes. My favorite Northwest deluge story comes from the Okanogan tribe.

Again there is the theme of a sadly disappointed divinity, a "tall white woman called Scomalt . . . she could create whatever she wished." Scomalt ruled over an island race of white giants forever at war among themselves. At last, Scomalt drove the wicked warriors to one end of White Man's Island, broke off that part of the land, and sent it far out to sea. What happened? All perished except that one good man and one good woman. The man catches a whale, stows its blubber in a hand-built canoe, and paddles almost forever until they come to the Okanogan country (upstate Washington). The legend ends with a warning:

> In time to come, the Okanogan Indians say, the lakes will melt the foundations of the world, and the rivers will cut the world loose. Then it will float as the island did many suns and snows ago. That will be the end of the world.

To what do all these many deluge myths refer? Could it be that myths are really memories? If, as the legends say, the world was once destroyed by water, wouldn't it be so catastrophic an event that its memory must survive? Whether or not the flood actually happened, what does it mean that our stories tell us that once we all drowned?

My father, the scientist, might say we're simply remembering our amphibious evolution as a species. My little sister, still a staunch Southern Baptist, might say it's God's punishment and get ready for the fire next time. I like to believe that we all have some memory of drowning because we are creatures who can live in two elements: the very earth and water that make up our bodies, as well as our world. And when I die, if given a choice, I'd just as soon stay here on Earth and undergo another sea change.

Once a psychic astrologer startled me by speaking about "my last death."

"Neptune . . ." she said softly. "You were lost at sea."

Whether I believed it or not, it felt true — or perhaps she was simply picking up on a two-year-old's undertow.

"Maybe . . ." I suggested hesitantly. "Maybe I was a whale?"

The large woman looked at me for a moment and I felt like a child again telling my sea stories when I should have kept silent. Then she laughed heartily and shrugged, "Well, maybe you were!"

And maybe I will be a whale again. I wonder what the world would be like if we told our children that their grandparents and other beloved, lost relatives are still living alongside us, that as the Northwest Coast Indians taught, the sea is the saltwater blood of our ancestors? If our myths and our bodies never abandoned the Earth for the sky, would we hold our natural world more sacred?

Of all the Northwest Coast Indian sea stories, the one that speaks most to me and restores my childhood faith is told in a two-volume story, *Renewal*, by Gua Gua La (Barbara Smith), a Cherokee woman living in Canada. In retelling the myth of those ancient mer-people, the Anishoni, and the prophesied reunion with their long-lost kin, the land-dwelling Ticanishoni, Gua Gua La speaks of a time we might all remember, a time before the waters and the People were divided.

This was long before Tyowa, the sun and singer of great songs, shone down so passionately upon the sea who bore his children, that the waters receded to expose Earth as well as some of the People of the Sea. They "now found themselves earthbound, took their bone knives and split open their tails, making for themselves legs and feet on which to walk the new land."

But all was not well with the People of the Earth, who forgot their mutual beginning in the womb of the sea and claimed a separate creation for themselves. No longer were the birds in the air or the fish of the sea their brothers. They saw these creatures as beneath them, made for their using. They forgot the universal speech which travels from mind to mind, linking all life. They stood alone on a high hill in the darkness, hearing and seeing nothing of the world around them, and for the first time in the world there was loneliness.

Loneliness, the myths tell us, always comes from forgetting who we are. Does the sea make us so lonely because we've lost our connection and come keening to her shore like motherless children?

Sometimes at night when I lie in my bed listening to the siren voice of a south wind singing off Puget Sound, when my bed rocks with its waves of air and outside the high tide crashes over its breakwall in a steady, sonorous shush like the sea herself is breathing, I close my eyes.

I'm lying on the very bottom of the ocean, my luminous, languorous tail as ballast and anchor. Graceful tentacles of kelp and the sponges' spiky tendrils wave in reunion. This underwater world holds one rhythm, holds all in its sea sway. I gaze up through an undulating ballet of tulip-shaped yellow stalks to see the shy shadow of an octopus, the white belly of a beluga whale, the darting cloud of a million minnows.

Far above is the sun, its warmth shining down like memory; below, pulled by the moon, the water glows a calm blue-green. Everywhere in these sea caverns are the tribes of my people. They're meditating, too, dreaming of a world far above them where they sometimes send their souls. They believe that we, their brothers and sisters of the air, have our own heaven up here — just as we long for their watery underworld. Maybe fathoms deep in the sea is where all the old and the new souls are dreaming and changing and being born again.

Living by Water

*W*ater carried me here; water keeps me here. I have most always lived by water — from Georgia's Yellow River to New York City's Hudson, whose winter ice floes crash with the sound of worlds colliding, to a wide irrigation ditch in arid Colorado, and now to this long embrace of Seattle's Puget Sound. Watching the water from my own window is so much second nature to me that in 1981, when I found myself staring at an Arizona sun boiling up over the horizon, heat glazing the low, red rocks like a mirage, I realized I had to do more than a rain dance for relief.

In a decision that capped my father's conviction that his eldest daughter was financially feebleminded, I sold my birthright to escape Arizona and move to Seattle. To do so, I traded in my precious and only inheritance: two water shares from my great-uncle's farm irrigation ditch company. My father didn't see the symbolism, that water was my medium of moving from desert to Northwest oasis. Now Father does note rather grudgingly that after nine years, my mutiny might have brought me more security than money because water seems to have at last settled me here.

"Water has more life in it than any part of the earth's surface," wrote Thoreau. With 80 percent of our bodies liquid, water is literally life to us. Water is also the first element we experience. We spend our formative nine months afloat in an amniotic sea so rich it re-creates the primal ocean as we move again through all stages of evolution: from reptilian fetal tail to amphibious gills and at last to lungs. We have the memory of nurturance flowing into our bellies through that fluid umbilical — a liquid lifeline.

However could we forget this first watery bond, even beached as we become in bodies that struggle against gravity, breathe high, harsh, and fast air, and at last lie down in solid ground? I suspect those of us who still feel the deep draw of water — whether we're Northwest fishermen, Seattle houseboat floaters, islanders, weekend sailing skippers, river rats; whether we live by Lake Washington or Union, Puget Sound, the Duwamish or eagle-haunted Skagit rivers, or the fertile Strait of Juan de Fuca — are people who feel our true roots not in soil, but sea. In fact, our psychic roots might look more like the delicate float of jellyfish than grounded tree ganglia.

What is it about the element of water that shapes and

characterizes our Northwest life? First, water is supportive. We've not yet worn it out as we have so much of our Earth. Water is also mutable; we cannot divide and determine it as we do our property. And by its very nature, water cannot be possessed — it flows right on through our hands.

To be supported daily by so yielding and yet strong a force as water perhaps lends the Northwest some natural grace that we would do well to consciously imitate. Compare a ferry full of commuters with a subway train tunneling through Manhattan stone. The first sails and rocks along with its lulled riders, while the latter jolts, thrusts, and speeds its passengers full tilt toward home. When I lived in New York City, I would never seek solace in a subway. But it is often to ferries I take my weary, sad, or forsaken self, as if the simple act of being on water will solve or erase. One of my friends stood on the slick deck of the Bainbridge ferry during a rainstorm and threw her wedding ring into the Sound. Only then did Deborah feel her divorce was final. Another friend got married on a ferryboat in the fog, those eerie foghorns sounding like the mournful celebration of whale songs.

Throughout history, water has offered this ceremonial certainty to our private and public rites. What is the first chore we do when someone is born or dies? We wash them. By such ablutions we baptize our comings and goings and remind ourselves that water is holy. It also has the power to heal.

In both our Western and Eastern religious traditions, water can purify, easing away our sins like so much silt and making of our souls a "new heaven and new earth."

Water is a world-changer. That's why in almost every mythology, it was water that once destroyed the world in a great flood. According to the Skagits, this great flood covered all except Kobah (Mount Baker) and Takobah (Mount Rainier). One of my favorite Northwest Coast

Indian deluge stories is the Vancouver Island *Daughters of Copper Woman*, by Anne Cameron (Press Gang Publishers, 1981), in which a solitary, sobbing Copper Woman cries all her tribe into creation.

If water is the original stuff of life, it's also a way of living. The ancient Chinese philosophy of Taoism, often called "the watercourse way," saw water as the most revered teacher. In the writings of Lao-tzu, the sixth-century B.C. master whose *Tao te ching* is still a classic:

> The highest good is like water,
> for the good of water is that it nourishes everything
>   without striving.
> The most gentle thing in the world overrides the most
>   hard.
> How do coves and oceans become kings of a hundred
>   rivers?
> Because they are good at keeping low —
> That is how they are kings of the hundred rivers.
> Nothing in the world is weaker than water,
> But it has no better in overcoming the hard.

To the Taoists, water's intuitive, harmonious flow was wiser than the rational attitudes of a linear, man-over-nature mind. Thus, according to the ancient Taoist Kuan-tzu, "the solution for the Sage who would transform the world lies in water." Our present-day willfulness has taught us much about our own separate selves and little about how to live wisely and at one with the world. Only now are we recalling that there is a Tao at play in the world; and understanding its way may well help us find reunion with our own, and the Earth's, body.

Here in the Northwest, where so much of our life is linked with water, where the Asian influence is keenly felt, we have a chance to follow this Tao or water's way. Perhaps that's why in the Northwest the most recent

Stephen Mitchell translation of the *Tao te ching* is a best seller — especially during the long, gray rains of winter. What more ideal landscape (except China itself) to learn that, as Mitchell explains of Lao-tzu's sage, "the Master has mastered Nature; not in the sense of conquering it, but of becoming it."

I'm convinced that at some point those of us who stay in the Northwest do *become* one with our inward, watery winters. We may rant about the rain, but if there are too many days of straight sun, an odd thing happens to true Northwesterners: we feel exposed and quite disoriented. We have more automobile accidents, we get exhausted from bearing up under a brilliant sun's mandate to be outdoors and too outwardly active. At last, in those stunning summer stretches, we get overstimulated by sunshine and seek the shade. With too much light, we seem to lose our life, seeking, in Lao-tzu's words, a more meditative way that is "radiant, but easy on the eyes."

My first awareness of this underlying way, or Tao of things, was while learning to drive a stick shift in the Virginia countryside. My father taught me to master the clutch and to shift gears, "with authority." My mother advised me to coax the recalcitrant shift through its lugubrious gears. But one day, grinding my way into second, an old backwoods relative who had never driven a day in her life leaned forward from the backseat and suggested rather shyly, "Seems to me that stick *wants* to go into its next gear, don't it? Natural-like. Because ain't that where it belongs now?" Only then did I understand: I was not master of the machine, nor slave to its whims. I was an integral part of its workings — its way.

A decade later I stood on the other side of the country, knee-deep in Colorado mud, working a farm my family inherited. I stood amazed to see the strong gush of irrigation ditch water so mildly disciplined into narrow ditches flowing through my cornfield. The water moved humbly along my elaborate system of wooden gates and hoed rows. I worked with my shovel, turning the water easily this way and that. By midnight, when the moon made the rivulets into quicksilver veins pulsing through my garden, I saw the luminous arteries of the Earth revealed. And though I was supposedly directing the flow, I felt no power. I felt peaceful. And there was something else — awe at the subtlety of this water whose strength was in surrendering to another element, earth.

These are two times in my life I've felt the harmony of letting go and allowing another force to move through me. Most recently, I've had this sense while staring at Puget Sound. Beach life here is elemental. My first day on the beach, I was shocked to find a great blue heron dying in the gentle surf that lapped my bare legs. Blood pooled in its beak, its eye already plucked out by a predator, as the bird sank into the sand, its old neck broken. I carried the heron up from the water and buried its body deep in the forest soil. Overhead gulls cried and I placed sand dollars as gravestone markers. Even then, in death, I felt a naturalness, the rightful end of a cycle.

Since that day years ago, I've apprenticed myself to Puget Sound because I believe it will teach me more about living than what I've learned so far. Maybe I hope its watery wisdom will seep into me every night as I lie, listening to the whoosh of waves against my seawall. Maybe I'd like to hook up again consciously to the umbilical that first nurtured me when I breathed water.

Whatever the reason, I've found solace in living by water. Solace and a sense of humor restored. It amuses me to think that waterfront property rights ebb and flow with the tides. Today my property might extend well beyond my neighbor's buoy with a minus 2 low tide; tomorrow my property shrinks back to a plus 3 tide. It is water that shapes and defines my boundaries, not the other way round.

In learning to yield to water's having its way with us, we change our character. The Northwest Indians were not as warlike as the prairie or desert tribes. They were fishermen and whalers following the waters, not warriors claiming the land. Water, contrary to even our West's labyrinthian water laws, cannot be truly claimed. It's too malleable. Water may reflect us, mirror our deepest selves, but it won't bear our imprint, our scars. We can in a sense wound it through pollution and our contempt, but more often than not it will wash away those wounds over time.

Water leaves no trace of us, though over eons it has left its own mark in the whorls of canyon stone, wide-wandering tributaries, glacial melt. More than volcanic fire and wind, earth is sculpted by water. Perhaps, then, it is from water we can at last learn how to shape ourselves in its image. And it is also from water that our scientists may now discover a future source of fusion — hydropower as an alternative form of energy or fuel.

Alan Watts, in his commentary on the *Tao te ching*, says, "The art of life is more like navigation than warfare, for what is important is to understand the winds, the tides, the currents, the seasons, and the principles of growth and decay, so that one's actions may use them and not fight them."

That first day on Puget Sound, I made a mistake in burying that blue heron. These years living by water have reminded me of what the Indians call the great giveaway of death. In the sea, all that dies sinks and rejoins the food chain. If one of the signs of our humanness was our Neanderthal Man burying his dead for the beyond, then perhaps one of our first signs of returning at last to the wisdom of the sea is to not always return dust to dust, but sometimes water to water.

So here where I have made my home, I am also making

my own way, my Tao. Here in the Northwest, I at last claim my home, my resting place. And if it is my choice when I die, then let me go back to the sea. Slip my body into the waters of this chosen land and let me feed the birds as I do now from shore. Let my body live and let me die, by water.

About the Author

A transplanted Southerner and graduate of the University of California at Davis, Brenda Peterson has worked for *The New Yorker*, lived on a farm near Denver, where she was a fiction editor for *Rocky Mountain Magazine*, and taught at Arizona State University in Tempe. For the past seven years she was an editor and environmental writer and has made her home in Seattle. Her first novel, *River of Light*, was published in cloth by Alfred A. Knopf in 1978, and reissued in paperback by Graywolf Press in 1986. Her second novel, *Becoming the Enemy*, was published by Graywolf Press in 1988, and her third novel, *Duck and Cover*, is forthcoming from Harper & Row in Spring 1991.

Many other fascinating books are available from Alaska Northwest Books™. Ask at your favorite bookstore or write us for a free catalog.

Alaska Northwest Books™

A division of GTE Discovery Publications, Inc.
P.O. Box 3007
Bothell, Washington 98041-3007
Call toll free 1-800-331-3510